In Memoriam—Charles Lutwidge Dodgson
1832–1898

In Memoriam
Charles Lutwidge Dodgson
1832–1898

Obituaries of Lewis Carroll
And Related Pieces

Compiled and Edited by

August A. Imholtz, Jr. & Charlie Lovett

New York • Lewis Carroll Society of North America

January 14, 1998

ISBN 0-930326-12-1

Library of Congress Catalog Card Number 97-075642.

Designed by Charlie Lovett.

Manufactured in the United States of America.

Table of Contents

Acknowledgements

The editors are grateful to the following individuals for assistance in locating obituaries and other fugitive literature: the late Joseph A. Brabant, Joel M. Birenbaum, Mark Burstein, Deborah Caputo, Morton N. Cohen, Selwyn H. Goodacre, Alan Tannenbaum, Bob Taylor, Alexander D. Wainwright, and Edward Wakeling. Thanks also to the staffs of the British Library, Newspaper Division; Guildford Muniment Room; The Library of Congress; Princeton University Library; and University of Maryland Library.

Note on Sources

Since only a few newspapers have published indexes to their 19th-century issues, the editors have examined column by column and page by page several hundred newspapers and journals from 1898. Newspapers were usually searched for approximately two weeks after Dodgson's death on January 14, 1898, though sometimes for a longer period. Journals, depending on their periodicity, were examined for the weeks or months following Dodgson's death; and in the case of many journals, tables of contents and annual cumulative indexes were routinely checked whenever possible.

American newspapers and journals, together with the more prominent British papers and journals, were located principally at the Library of Congress in Washington, D.C. and the University of Maryland Library. These titles, except for some of the bound periodicals, were read on microfilm. The provincial British papers and some other titles were consulted in hard copy and microfilm at the Newspaper Division of the British Library. Numerous British and Commonwealth items are preserved in a Dodgson family album at the Guildford Muniment Room. In some cases, these items do not include complete publication information, and obituaries reprinted here without dates or places of publication are from that album and could not be located in their original form. Some items were located in private collections, including those of the editors.

The editors are clearly aware that this volume contains only a selection of the obituaries and related notices on the death of Charles Lutwidge Dodgson; but it is hoped that these texts provide a representative appreciation of the popular culture's estimation of Lewis Carroll in Great Britain, the United States, and the Commonwealth.

Introduction
Charles Dodgson and the Year of his Death
Charlie Lovett

Charles Lutwidge Dodgson began 1898, as he did most years later in life, at the home of his sisters in Guildford, The Chestnuts. He had traveled there for the holidays on 23 December 1897 and seemed, as his nephew and biographer Stuart Dodgson Collingwood wrote, "to be in his ordinary health, and in the best of spirits." He was working hard on the second part of his *Symbolic Logic* (*Part I—Elementary* had already reached a fourth edition and *Part II—Advanced* and *Part III—Transcendental* had been announced as forthcoming). Dodgson made no entries in his diary during the new year, but we may assume that he passed the change of years quietly with his family, probably attending services at St. Mary's Church nearby on New Year's Day and on Sunday, January 2.

The following Wednesday, January 5, a telegram brought the news that Dodgson's brother-in-law, Charles Edward Stuart Collingwood, Rector of Southwick near Sunderland, had died. The family requested that Dodgson come at once. Dodgson wrote to his newly widowed sister Mary, "I would certainly have come to you, if I could have done so with reasonable prudence: but, with a feverish cold, of the bronchial type, and the risk of ague (a form my colds usually take), Dr. Gabb forbids me to risk it." He sent his sister £50 "on account" so that she should have some "ready money" and also wrote a letter to her son warning him about dealings with the undertaker. Of the tens of thousands of letters he had written in his life, these were the last two.

In his biography of his uncle, Stuart Dodgson Collingwood (that same nephew), takes up the story.

> At first his illness seemed a trifle, but before a week had passed bronchial symptoms had developed, and Dr. Gabb, the family physician, ordered him to keep to his bed. His breathing rapidly became hard and laborious, and he had to be propped up with pillows. A few days before his death he asked one of his sisters to read him that well-known hymn, every verse of which ends with "Thy Will be done." . . . He seemed to know that the struggle

> was over. "Take away these pillows," he said on the 13th, "I shall need them no more." The end came about half-past two on the afternoon of the 14th. One of his sisters was in the room at the time, and she only noticed that the hard breathing suddenly ceased.

January 14th was a Friday, less than a fortnight shy of Dodgson's sixty-sixth birthday. The next day, and in the days that followed, the papers were filled with accounts of his life. His secret identity as "Lewis Carroll," which he had so assiduously protected from revelation in the press, was now announced in papers around the world. And because Dodgson had shielded his private life from public connection with "Lewis Carroll," the obituaries which flooded the papers were the first glimpse that the world at large was afforded of the life of a beloved author—and the myths created as part of that "first impression" persist to this day.

The obituaries made much, for instance, of the dichotomy between the serious Rev. Charles L. Dodgson and the humorous Lewis Carroll. A close examination of the facts of his life shows that no such split personality existed, that Charles Dodgson and Lewis Carroll were the same man, with the same core of faith and values. Yet, the myth persists and is still often repeated both in print and, more insidiously, by word of mouth. The obituaries, in fact, prove to be a source for many misconceptions about Dodgson—that he had few adult friends or was only humorous with children, that he abruptly ended all his child friendships when the child reached puberty, that he was reclusive and antisocial—and these myths, by their repetition, have largely framed the general public's (that is the non-biography-reading public's) perception of Dodgson.

True the obituaries were full of factual errors—that Dodgson was born in Croft in 1833 or that he wrote books called *Euclid and the Game of Logic* and *Symbolic Life*—but these errors are easily corrected and do not, ultimately, affect our view of Dodgson as a man. It is rather those misinterpretations, based on the need to quickly generate an obituary with few accurate sources, which caused the greater damage. Perhaps, if Dodgson had been more open about his private life, obituary writers would have had more reliable sources. As it is, many of the obituaries seem to depend partially if not mostly on rumor, and once that rumor made it into print, it was copied by other papers as fact.

But the obituaries were not exclusively made up of myth and rumor. Certainly they informed the public of much of the basic information of Lewis Carroll's life, about which it had been largely ignorant. His work at Christ Church as a mathematician, his ability to enchant young girls, his passion for theatregoing—all these elements of truth are repeated in the obituaries. What was written about Dodgson at the time of his death was *mostly* true, making it all the more difficult to separate the few misconceptions that entered the public mind at that time and to tell the public now, a century later, "No! This part is wrong!" Lewis Carroll's obituaries brought Charles Dodgson to the world and they set that world on a path towards understanding the man, and misunderstanding him.

Dodgson's passing was noted not only in newspapers around the world. In the Cathedral at Christ Church, Oxford, two days after Dodgson's death, Canon William Sanday preached a sermon (quoted extensively in the press, see for instance, pp. 72–74) in which he remarked on Dodgson's life and works.

Two days later, on Tuesday January 18, a memorial peal was rung on the Christ Church bells. Ironically, the bells which rung this peal still hung in the wooden belfry which Dodgson had so derided in his humorous pamphlets on the architectural improvements at Christ Church. Though hidden by the stone bell-tower of G. F. Bodley, completed in 1879, the "meat-safe," as Dodgson called it, holds Christ Church's bells to this day. According to S. A. Warner's *Oxford Cathedral*, the peal of Kent Treble Bob Maximus rung on 18 January 1898 consisted of "5,040 changes . . . rung in three hours fifty-eight minutes, muffled in memory of the Rev. C. L. Dodgson." "It was a curious thing," writes Warner, "that, during the ringing of this peal, the Very Rev. H. G. Liddell, sometime Dean of Christ Church, passed to his rest in Ascot, Berkshire."

Liddell and Dodgson had been linked in life—sometime friends, on occasion adversaries in Christ Church and University business, and connected because of Dodgson's close friendship with the Dean's daughter Alice, for whom *Alice's Adventures Underground* was written. Now the two were linked in death. Some of the Oxford papers mentioned both men in the same columns, and when Francis Paget, then Dean of Christ Church, rose to the pulpit in the Cathedral the following Sunday, January 23, he preached a sermon, "The Virtue of Simplicity," in honor of both men (see below page 176–180).

Dodgson had left instructions that his funeral should be "simple and inexpensive, avoiding all things which are merely done for show, and retaining, only what is, in the judgement of those who arrange my Funeral, requisite for its decent and reverent performance." Dodgson had appointed his brothers Wilfred Longley Dodgson and Edwin Heron Dodgson as executors of his estate, but as Edwin was at the time vicar at Jamestown, St. Helena, in the South Atlantic, it was Wilfred who undertook to make the funeral arrangements in Guildford. Wilfred wrote to

Francis Paget, inviting him to assist with Dodgson's funeral. On January 17, the Dean replied:

> I am deeply grateful to you and to your sisters for the great kindness which your letter shows me. If all be well, I will gladly come straight to your house from the station on Wednesday morning: and I will bear in the service the part you bid me bear. I am very sensible of the trust you give me.

The funeral was held on Wednesday, January 19, at St. Mary's, Guildford, with the burial following on The Mount. Details of the service are given in articles from the local and national papers in the main text of this book. Dean Paget and the Rector of St. Mary's, Canon Grant officiated. Dodgson's name was duly entered in the St. Mary's Register of Burials as number 123, with the word "Priest" written below it in pencil. He was one of fifteen buried in the parish in 1898.

Dodgson was given perhaps the most visible site in the graveyard near the top of The Mount in Guildford. His brothers and sisters paid for the erection of a white cross surmounting three steps. On the cross are the words "THY WILL BE DONE," no doubt inspired by Dodgson's desire to hear the words of that hymn. Below, the engraving reads,

WHERE I AM THERE SHALL
ALSO MY SERVANT BE.
REVd. CHARLES LUTWIDGE DODGSON,
(LEWIS CARROLL,)
FELL ASLEEP JAN. 14, 1898,
AGED 65 YEARS.
HIS SERVANTS SHALL SERVE HIM.
"FATHER IN THY GRACIOUS KEEPING,
LEAVE WE NOW THY SERVANT SLEEPING."

On one side of the stone base is the inscription "ERECTED IN LOVING MEMORY BY HIS BROTHERS AND SISTERS."

On 4 November 1871, Thomas Vere Bayne and Augustus Charles Vernon Harcourt, both Students of Christ Church and friends of Dodgson, witnessed the will of their colleague, which read:

> THIS IS THE LAST WILL AND TESTAMENT of me CHARLES LUTWIDGE DODGSON, Student of Christ Church, Oxford. I give, devise, and bequeath all the real and personal

> estate of which I shall be possessed or entitled to at the time of my decease to be divided into equal shares, one share to go to each of my brothers and sisters who shall be then living, and if any one of them be then deceased but have married one share to be divided equally among the children, if any be then living, of such brother or sister. And I appoint my brothers Wilfred Longley Dodgson and Edwin Heron Dodgson EXECUTORS of this my will. . . .

Wilfred Dodgson began the task of settling his brother's estate almost immediately. He carefully recorded income flowing into the estate from Macmillan and other publishers of Dodgson's works, the sale of stocks, and the sale of Dodgson's personal assets. He paid outstanding bills, including charitable pledges (£135 to Oxford High School, for instance). Funeral expenses of £28 18s were paid out along with 8s 8d to print a memorial card which bore a photograph of Dodgson and an excerpt from his *Easter Greeting*. Because Dodgson had lived in college rooms at Christ Church, it was important to clear away his personal effects as quickly as possible so that the rooms might be turned over to a new occupant. Wilfred allowed each of the Dodgson brothers and sisters to purchase items from the estate that they wished to keep (he himself bought Dodgson's Hammond typewriter and a group of unlisted miscellaneous items for £6). A painting which had struck the fancy of the Rev. George William Kitchin, a former Christ Church Student with Dodgson and the Dean of Durham, was sold to him for £25.

Dodgson's set of rooms in the northwest corner of Tom Quad had not only to be stripped of their contents, but returned to their original condition, and this meant the removal of Dodgson's "glass house" photographic studio on the roof. Though Dodgson had not used the studio for photography for over 15 years, it was still there, until Wilfred paid Hughes and Symm £104 10s for "removal of studio and restoration of roof."

The vast majority of Dodgson's belongings remained to be disposed of—furniture, personal effects, and thousands of books. Wilfred engaged Mr. E. J. Brooks of Oxford to oversee the disposal and sale of these remaining items. The first order of business was to clear out the large quantity of personal papers. Because these papers were destroyed, we cannot be sure exactly what remained at the time of Dodgson's death. Some mathematical manuscripts did survive, but Dodgson's letter register, his letters, his personal files, and who knows what else, were all destroyed. On February 10, Wilfred wrote to Brooks,

> I thought at the finish that it would be a great deal of trouble bringing the sacks of papers all the way down here, and, as you stated that you could have them burnt in the manner we wish, I should be glad if you would do so.

After burning the papers, Brooks turned his attention to the remaining items. The majority of these were sold at an auction at the Holywell Music Rooms in Oxford on Tuesday and Wednesday, May 10–11, 1898. The printed catalogue for that sale has been reproduced in Jeffrey Stern's *Lewis Carroll Bibliophile*. The *Oxford Times* wrote that "there was a large attendance of dealers, members of the University, and citizens at the sale." After Brooks' commission, the sale netted the estate £902 2s 3d. Several of the items from the sale, including first editions of *Alice's Adventures in Wonderland*, would today be worth hundreds of thousands of pounds.

The final settlement of the estate had waited only for the income from the Brooks sale. On May 13, Dodgson's will was "proved and registered in the Principal Probate Registry of Her Majesty's High Court of Justice." The gross value of the estate was stated as £4145 11s 1d and £118 0s 7d had been paid in Estate Duty. When all the bills had been paid, £4004 9s 4d remained to be divided among Dodgson's brothers and sisters, all ten of whom survived him.

While the settlement of Dodgson's estate, except for the Brooks sale, was a relatively private matter, "Lewis Carroll" continued to be in the public eye. As the months after his death wore on, more reminiscences appeared in the press. The present volume is limited to pieces published within about six weeks of January 14, in order to show how Dodgson was viewed at the time of his death, but Morton Cohen's collection *Lewis Carroll: Interviews and Recollections*, includes many of these later narratives.

Lewis Carroll's final book, *Three Sunsets and Other Poems* was published in February 1898. Most of the poems were previously published, but Dodgson had commissioned his friend Gertrude Thomson (whose reminiscence of him appears below, pp. 53–64) to create a series of fairy drawings to accompany his poems.

Efforts were also underway to establish a permanent memorial to Dodgson. On February 15, 1898, an article appeared in the *St. James's Gazette*, a newspaper to which Dodgson had once been a regular contributor, under the heading "For the Sick Little Ones—A Lewis Carroll Memorial In the Children's Hospital." "We are only too ready," it read, "to respond to the request of many influential friends, and to start a Fund by which the desire to give shape to the gratitude of countless lovers of this lover of children may be satisfied." The article proposed endowing a cot in the Children's Hospital in Great Ormond Street at the cost of £1000 and listed a "General Committee" for the fund which included Alice Hargreaves, John Tenniel, Henry Irving, and other friends of Dodgson and public figures (see below pp. 183–84n for a complete list of the Committee). Similar cots had already been established at the hospital in memory of Margaret Scott Gatty, author of the popular children's book *Parables from Nature* and editor of *Aunt Judy's Magazine* (to which Dodgson had been a contributor); and Julia Horatio Ewing, author of *Jackanapes* and other children's books

The following day an article by the poet Alice Meynell appeared in the *St. James's Gazette* under the title "The 'Lewis Carroll' Cot" (complete text on pp. 180–83). "The first suggestion that the author of *Alice in Wonderland* should be commemorated by the endowment of a cot," wrote Meynell, ". . . was made by a child, a little friend of Lewis Carroll's and the daughter of one who was in her childhood his little friend also." Unfortunately, the article does not give the identity of the child who suggested this memorial.

A few days later a circular letter (text below on pp. 183–85) was printed and distributed giving the names of the General Committee and a list of those who had already subscribed. Over the following weeks, the *St. James's Gazette* published frequent updates on the progress of the fund and printed the names of contributors and the amounts contributed. The lists are filled with noteworthy Victorians, as well as personal friends of Dodgson, as a representative sampling will show:

From the theatre world: Edgar Bruce, who helped produce the first full stage version of *Alice* in 1886 (£5); Sir Henry Irving (£5); Irene Vanbrugh, who had been a child-friend (£1 1s); Mrs. Arthur Lewis, the former Kate Terry, whose entire family had been close to Dodgson (£1 1s); J. L. Toole, the actor whom Dodgson had seen on many occasions on stage (£1 1s); and Mrs. Frank Gielgud, niece of Ellen Terry and mother of John Gielgud (£1).

From the art world: John Tenniel, illustrator of the *Alice* books (£5 5s); Henry Holiday, illustrator of the *Snark* (£1 1s); Sir Noël Paton, whom Dodgson had approached about illustrating *Looking-Glass* (£1 1s); William Holman Hunt, the Pre-Raphaelite painter (£1 1s); Walter Crane, Arts and Crafts designer and children's book illustrator (£2 2s); and Philip Burne-Jones, son of Edward, the Pre-Raphaelite Painter (£1 1s).

From his colleagues and former colleagues at Oxford: Canon Robinson Duckworth, the other adult present on the famous river journey of 1862 (£2 2s); Sir Henry Longley, who had been at both Rugby and Christ Church with Dodgson (£3); Sir Michael Hicks-Beach, sometime Chancellor of the Exchequer (£3 3s); Canon Sanday who had mentioned Dodgson in his Cathedral sermon (£3 3s) and the Canon's wife (10s); Rev. Canon Jelf (10s); and the Bishop of Gibralter (£1 1s).

From the Royalty and aristocracy: The Princess of Wales (£10); The Duke of Marlborough, first cousin to Winston Churchill (£1); The Duke and Duchess of York (£5 5s); The Duke of Norfolk (£5); The Duke and Duchess of Fife, the Duke being the President of the Hospital for Sick Children (£20); Princess Henry of Prussia (£2 2s); the Duke and Duchess of Bedford (£10); the Duchess of Albany was a contributor, and Princess Alice of Albany herself even"subscribed a guinea from her own purse."

From the literary world: Jerome K. Jerome, author of another famous "floating on the Thames" story, *Three Men in a Boat* (£1, 1s); W. M. Rossetti, the brother of the late poet and painter Dante Gabriel Rossetti—Dodgson had photographed both men—(£1 10s); George MacDonald, fantasy author of *At the Back of the North Wind* and a friend of Dodgson (£2 2s); and Austin Dobson, who, in 1907, would write a "Proem" to the Arthur Rackham edition of *Alice*.

His own family member supported the fund generously: The brothers and sisters of Lewis Carroll (£21); C. Fletcher Lutwidge (£5 5s); Stuart Dodgson Collingwood, his nephew and biographer, was one of several who cleverly contributed the price of the Hatter's hat (10s 6d); and various members of the Wilcox family contributed a total of £2 14s.

And other familiar names: Mrs. Andrew Carnegie, wife of the famous philanthropist (£1 1s); Messrs. Macmillan and Company, publishers of many of Dodgson's works, including the *Alice* books (£21); and Charles Manners, who had, with Dodgson's permission, produced a Looking-Glass Biscuit Tin in 1892 (10s 6s).

From those he had known as children: Alice Hargreaves (£10 10s); Greville MacDonald, son of George MacDonald, and the child who had said, on reading the manuscript of *Alice*, that he wished there were sixty-thousand volumes of it (£1 1s); Mrs. Lorina Skene, formerly Lorina Liddell, (£1 1s); Beatrice Hatch (£1 1s); Agnes Weld (£1 1s); Enid Stevens (£1 1s); Gertrude Chataway, to whom Dodgson dedicated the *Snark* (10s); and Alexandra Cardew, the former Xie Kitchin (£1 1s).

Dodgson would doubtless have been most pleased by the many contributions from groups of schoolchildren which were received, such as: Lewisham High School for Girls (£1 8s); Students of Somerville College, Oxford (£3 4s 9d); Worcester High School for Girls (£1 1s); The

Girls of the Manor House School, Dunstable (10s); The Pupils of Lincoln House, Harrow (5s); St. Hugh's Hall, Oxford (10s 9d); Girls High School, Leeds (£2); Girls High School, Oxford, where Dodgson had lectured on Logic, (£4 0s 7d); and many other school groups.

Of course, there were many anonymous gifts as well, often coming under the guise of a Carrollian name. Contributions were listed from Bill the Lizard, The Cheshire Cat, The Treacle Well, The Duchess's Baby, The Slithy Toves, The Mock Turtle, and others.

These names are merely representative—many others donated as well, from members of the aristocracy, to adult friends of Dodgson, from those who had been child-friends, to strangers who only knew him through the *Alice* books.

Some contributions came with accompanying letters which were quoted in the *St. James's Gazette*. Alice Hargreaves wrote:

> Most of Mr. Dodgson's stories were told to us on river expeditions to Nuneham or Godstow, near Oxford. My eldest sister, now Mrs. Skene, was 'Prima,' mentioned in the poem at the beginning of *Alice's Adventures in Wonderland.* I was 'Secunda,' and 'Tertia' was my sister Edith. I believe the beginning of 'Alice' was told one summer afternoon when the sun was so burning that we had landed in the meadows down the river, deserting the boat to take refuge in the only bit of shade to be found, which was under a new-made hayrick. Here from all three came the old petition of 'Tell us a story'—and so began the ever-delightful tale. Sometimes to tease us—and perhaps being really tired—Mr. Dodgson would stop suddenly and say, 'And that's all till next time.' 'Ah, but it is next time,' would be the exclamation from all three; and after some persuasion the story would start afresh. Another day, perhaps, the story would begin in the boat, and Mr. Dodgson, in the middle of telling a thrilling adventure, would pretend to go fast asleep to our great dismay.
>
> I have often thought, with gratitude and wonder, of the unvarying kindness and good nature shown to us. Alice's adventures were first written down in answer to my teasing wish to possess the story in book form. Long before this I can remember, in the intervals between photographing one or other of us, Mr. Dodgson would make quaint drawings in the style of "Father William," and fit them to a story he told as the drawings were made. Sheets of these were treasured by me, but, alas! none remain now.

This, incidentally, was the first published account of the genesis of *Alice* written by Mrs. Hargreaves, who had additional reminiscences published in 1932 at the time of Dodgson's centennial. It was published in part in Collingwood's, *The Life and Letters of Lewis Carroll*, but the final paragraph was omitted.

Henry Irving wrote:

> Any work associated with the name of Lewis Carroll must command wide-spread interest and sympathy, more especially anything which is connected with the care and happiness of children . . . There could be no more fitting memorial of Lewis Carroll than the cot in the Children's Hospital. I sincerely trust this scheme will receive a most encouraging measure of public support.

And Sir Walter Besant:

> In sending you a small contribution to your Cot from my daughters, I take this opportunity of congratulating you on the work you have undertaken. It is now thirty years since I first made the acquaintance of *Alice*, and I should not like to count the number of copies of that incomparable work which I have given to little girls of my acquaintance since that time. It is the only child's book, perhaps, which can be read with equal pleasure by old and young. I should frequently read it still but for the fact that I know it by heart. It is also the only child's book of nonsense which is never childish, although it always appeals to a child; where there is no writing down to the understanding of a child, though it can always be understood by a child. It is, in a word, a book of that extremely rare kind which will belong to all generations to come, until language becomes obsolete. I cannot doubt that your cot will be more than provided. If there is any doubt upon the subject, you have only to ask us all to double our subscriptions.

By April 28 the *St. James's Gazette* could announce that the £1000 had been raised, and contributions continued to flow in. In November of 1898 a formal presentation of the funds was made to endow the cot.

At the same time, Stuart Dodgson Collingwood was at work on what would be the other lasting memorial to Dodgson to emerge during 1898—his biography of his uncle. On August 22, *The Times* of London carried the following item, under the heading "'Lewis Carroll's' Letters":

> Mr. Stuart D. Collingwood writes from the Chestnuts, Guildford:—"I shall be greatly obliged to you if you will allow me to appeal through your columns to readers of *The Times* who may have had any personal acquaintance with the late Rev. C. L. Dodgson. I am very anxious to obtain copies of such letters of his as would be interesting to the public, and also any reminiscences of him, anecdotes about him, &c., suitable to be included in the memoir of his life which I am writing. . . .

The Life and Letters of Lewis Carroll was published by T. Fisher Unwin, London in December of 1898 and by The Century Company in New York. Excerpts appeared in the December issues of *The Strand Magazine* and *The Century Magazine*, and an interview with Collingwood appeared in the December 9 issue of *The Westminster Budget*. Reviews of the book concentrated more on summarizing Dodgson's life or retelling the various anecdotes which the book included, rather than evaluating the quality of the biography itself. The critic for *The Bookman* (London) wrote:

> We have read this biography with the utmost care, and are as far as ever from any real understanding of the man—unless it be understanding him to admit him an enigma. . . . It is an affectionate biography, and affection must always be partial. It is all the pleasanter for that, and very likely all the truer.

Though Collingwood's affection for his uncle and his surviving family may have made for a partial biography, there is no doubt that his *Life and Letters* has been a key source for every biographer who has followed in his footsteps, and over a dozen have done so, with more certainly to come.

By the time of Dodgson's death the place of Lewis Carroll as the author of *Alice's Adventures in Wonderland* and *Through the Looking-Glass and What Alice Found There* was secure in history. There may have been some question, however, about the extent to which the life of Charles Lutwidge Dodgson would elicit interest beyond those two pseudonymous volumes. But in the year following his death, the pattern of studying, remembering, eulogizing, analyzing, and critiquing the life of Charles Dodgson became established. From the depth of the longer obituaries to the outpouring from the people in support of the Lewis Carroll

Cot; from the publication of lengthy reminiscences to the success of Collingwood's first biography—the world desired, as it still does, to learn about Dodgson, and though that desire may have begun with a fascination with the *Alice* books, it quickly extended far beyond that.

In 1899, Dodgson's former child-friend Isa Bowman would publish her *The Story of Lewis Carroll* and S. D. Collingwood would issue a collection of lesser known Dodgson pieces under the title *The Lewis Carroll Picture Book.* The appetite for Dodgsoniana had been established almost as soon as "Lewis Carroll" was dead. And still it grows.

A Caution to Readers

Obituaries are generally written under the pressure of a deadline with few accurate sources. The pieces in this book are filled with factual errors about the life and works of Charles Dodgson. Many of these errors are identified in footnotes, but in order to prevent the notes from overwhelming the text, we have not noted mistakes in the titles and dates of publications of Dodgson's works nor those in the dating of significant events in his life. Instead, we include here a list of the correct titles and publication dates of the works mentioned in the text as well as a brief chronology of Dodgson's life. We ask that the reader refer to these when questions of accuracy arise.

The obituaries have been transcribed as they appeared, including incorrect spellings, but some slight changes in typography have been made. Some passages are repeated from one obituary to another, and in many cases these repetitions have been noted. All obituaries included here are reprinted in their entirety, however, so that those wishing to see how Dodgson was remembered in a particular paper may do so.

Works of C. L. Dodgson

A Syllabus of Plane Algebraical Geometry (1860)
The Formulæ of Plane Trigonometry (1861)
A Guide to the Mathematical Student in Reading, Reviewing, and Writing Examples (1864)
The New Method of Evaluation as applied to π (1865)
Alice's Adventures in Wonderland (1865)
An Elementary Treatise on Determinants (1867)
Offer of the Clarendon Trustees (1868)
Phantasmagoria and other Poems (1869)
Songs from 'Alice's Adventures in Wonderland' (with some new words by Dodgson, 1870)
Through the Looking-glass and What Alice Found There (1871, dated 1872)
The New Belfry of Christ Church, Oxford (1872)
Facts, Figures, and Fancies Relating to the Elections to the Hebdomadal Council (1874)

Euclid, Book V., Proved Algebraically (1874)
The Hunting of the Snark: An Agony in Eight Fits (1876)
An Easter Greeting to Every Child Who Loves Alice (1876)
Memoria Technica (1877)
Euclid and his Modern Rivals (1879)
Doublets: A Word Puzzle (1879)
Lanrick—A Game for Two Players (first published as *Lanrick* 1881)
Mischmasch (1882)
Euclid, Books I and II (1882)
Lawn-Tennis Tournaments (1883)
Rhyme? and Reason ? (1883)
Christmas Greetings (1884)
A Tangled Tale (1885)
The Game of Logic (1886)
Alice's Adventures Underground (1886)
Curiosa Mathematica Part I A New Theory of Parallels, (1888)
The Nursery Alice (1889)
Sylvie and Bruno (1889)
Curiosa Mathematica Part II Pillow Problems Thought Out During Sleepless Nights (1893)
Sylvie and Bruno Concluded (1893)
Symbolic Logic Part I Elementary (1896)

Chronology of the Life of C. L. Dodgson

1832 Born in Daresbury, Cheshire
1843 Moves to Croft, North Yorkshire
1844–45 Attends Richmond Grammar School
1846–49 Attends Rugby School
1850 Matriculates at Christ Church, Oxford
1852 Becomes Student of Christ Church
1854 Receives B.A.
1855–81 Mathematical Lecturer, Christ Church
1857 Receives M.A.
1858 Becomes Senior Student under Ordinance of 1858
1861 Ordained deacon
1863 Serves as Mathematical Examiner
1868 Death of Father, family moves to Guildford
1898 Dies at Guildford

In Memory Of
Maxine G. Schaefer
(1925–1996)

The Academy (London)[1]
January 22, 1898

A methodical correspondent who has a passion for preserving literary odds and ends, and a capacity for finding them when needed, sends us the following advertisement. It was forwarded wholesale through the post by Lewis Carroll at the end of the year 1893:[2]

> For over twenty-five years I have made it my chief object, with regard to my books, that they should be of the best workmanship attainable for the price. And I am deeply annoyed to find that the last issue of *Through the Looking-Glass*, consisting of the Sixtieth Thousand, has been put on sale without its being noticed that most of the pictures have failed so much in the printing as to make the book not worth buying. I request all holders of copies to send them to Messrs. Macmillan & Co., 29, Bedford-street, Covent Garden, with their names and addresses; and copies of the next issue shall be sent them in exchange.
>
> Instead, however, of destroying the unsold copies, I propose to utilise them by giving them away to mechanics' institutes, village reading-rooms, and similar institutions where the means for purchasing such books are scanty. Accordingly, I invite applications for such gifts, addressed to me, 'care of Messrs. Macmillan.' Every such application should be signed by some responsible person, and should state how far they are able to buy books for themselves, and what is the average number of readers.

---

[1] This article was reprinted, with the omission of a few paragraphs, in the March 1898 edition of *Current Literature* (New York).

[2] Copies of this advertisement sheet were issued primarily by their insertion into copies of the first edition of *Sylvie and Bruno Concluded*. It was also printed as an advertisement in *The Times* on 2 December 1893.

> I take this opportunity of announcing that, if at any future time I should wish to communicate anything to my readers, I will do so by advertising in the 'agony' column of some of the daily papers on the first Tuesday in the month.
>
> Lewis Carroll.
> Christmas, 1893.

A Correspondent writes: "I read with interest your 'Book Reviews Reviewed' columns every week; and it may interest some of your readers to know how *Alice in Wonderland* was received on its first appearance.[1] I cannot discover that its merits were fully perceived, or its success predicted, by any critic. *The Times*, reviewing the book among a dozen other Christmas books, gave high praise to Mr. Tenniel's[2] drawings, but concerning Lewis Carroll's text only remarked that it was 'an excellent piece of nonsense.' *The Spectator* did not, I think, review the book at all on its first appearance. *The Athenaeum* indulged, of course, in a little dragon-slaying:

> This is a dream-story; but who can, in cold blood, manufacture a dream with all its loops and ties, and loose threads, and entanglements, and inconsistencies, and passages which lead to nothing, at the end of which Sleep's most diligent pilgrim never arrives. Mr. Carroll has laboured hard to heap together strange adventures and heterogeneous combinations, and we acknowledge the hard labour. Mr. Tenniel, again, is square and grim and uncouth in his illustrations, howbeit clever, even sometimes to the verge of grandeur, as is the artist's habit. We fancy that any real child might be more puzzled than enchanted by this stiff, overwrought story.

In his sermon at Christ Church, Oxford, on Sunday morning, Canon Sanday[3], Margaret Professor of Divinity, referred to the death of Lewis Carroll. We quote a few words:

---

[1] A collection of original reviews of *Alice's Adventures in Wonderland* appears in *Jabberwocky—The Journal of the Lewis Carroll Society*, volume 9, numbers 1–4 (Winter 1979/1980–Autumn 1980). The reviews quoted here all appear in their entireties in volume 9, number 2 (Spring 1980).

[2] Sir John Tenniel (1820–1914), political cartoonist for *Punch* and illustrator of the *Alice* books.

[3] Rev. William Sanday (1843–1920), lecturer in Theology at Trinity College, Oxford

> Might they not say that from their courts at Christ Church there had flowed into the literature of their own time a rill bright and sparkling, healthgiving, and purifying wherever its waters extended? . . . They in that place knew how fully the man bore out the promise of his books. . . . They knew his fondness for children and what trouble he took to make them happier and better. But, most of all, they knew what was the fount and spring from which all these varied activities took their direction. They knew how behind them all there lay a deep background of religion—a religion severely quiet and retiring, like his character—a religion almost of the closet after the pattern of the Gospel.

Mr. Dodgson's own manner as a preacher was earnest and slow. For several years he delivered the New Year sermon at St. Mary's, Guildford.[1]

Lewis Carroll was as fortunate in his illustrations as any writer could be. Under any circumstances the *Alice* books would have won tremendous favour, but not a little of their popularity must, none the less, be due to Sir John Tenniel's drawings. Artist and author have rarely been in such perfect accord. Again, in *The Hunting of the Snark*, Mr. Henry Holiday[2] is the poet's very faithful and admirable ally, catching the spirit of the nonsense to perfection. His beaver, looking "unaccountably shy," is the prince of beavers. And in *Sylvie and Bruno*, Mr. Furniss[3] did some of his best work. One reason of this high level of excellence is undoubtedly Lewis Carroll's interest and desire to have everything quite "right," and according to his own ideas. Of no man may it more truly be said that until he was satisfied he was dissatisfied.

By one of those curious and not uncommon coincidences, Lewis Carroll's friend, the father of the original Alice—Dean Liddell—died within four days of the author of the *Wonderland* books.[4] Dean Liddell's

---

(1883), Fellow of Exeter College, Oxford (1883 and 1888), and the holder of numerous posts both within Oxford and elsewhere. Sanday was also Lady Margaret Professor of Divinity and became a Canon of Christ Church in 1895. Dodgson's published diaries record one meeting with Sanday in 1897.

[1] Though he often passed the new year at Guildford after he moved his family there following the death of his father in 1868, only twice did Dodgson record delivering the New Year sermon, in 1887 and 1897. From 1887–97 he preached there eight times.

[2] Henry Holiday (1839–1927), English painter, sculptor, illustrator and designer of stained glass windows. He illustrated *The Hunting of the Snark* (1876).

[3] Harry Furniss (1854–1935), illustrator of the *Sylvie & Bruno* books, artist for *Punch*.

[4] Henry George Liddell (1811–98) served as Dean of Christ Church from 1851 until 1891. His daughters Lorina (1849–1930), Alice (1852–1934), and Edith (1854–76) (par-

name will ever be associated with that of Dr. Scott,[1] Jowett's[2] predecessor as Master of Balliol, for their invaluable lexicon. The fame of Liddell and Scott is inextinguishable. It may not here be out of place to tell again an old story of Dean Liddell and an undergraduate. "What Sophocles do you know?" the Dean asked. "Oh, I know all Sophocles," was the answer. "Really! I wish I could say the same." The victim began to translate "Where did you get that from?" asked the Dean with reference to a "howler." "Oh, Liddell and Scott." "Then," said Liddell with much gravity, "it was Dr. Scott's doing and not mine."

The Academy (London)[3] January 22, 1898

If I have written anything to add to those stories of innocent and healthy amusement that are laid up in books for the children I love so well, it is surely something I may hope to look back upon without shame and sorrow (as how much of life must then be recalled!) when my turn comes to walk through the valley of shadows."

These words were written in 1876 by Lewis Carroll in *An Easter Greeting to Every Child that loves Alice.*[4] And now his turn has come. Truly, he had no cause to feel anything but satisfaction. The world can show few writers who from first to last have used their talents so joyously, diligently, and to such kindly purpose as Lewis Carroll.

Lewis Carroll's best period lasted, roughly, from his thirtieth to his forty-fifth year. He began *Alice's Adventures Underground* in July, 1862;[5]

---

ticularly Alice), were the inspiration for many of Dodgson's stories in the 1850s and 1860s, including *Alice's Adventures in Wonderland.* Liddell died on 18 January 1898.

[1] Robert Scott (1811–87), joint author with Henry George Liddell of the Liddell-Scott *Greek-English Lexicon.* Dean of Rochester from 1870 until his death.

[2] Benjamin Jowett (1817–93), Master of Balliol and Regius Professor of Greek. Dodgson became involved in the controversy over Jowett's annual salary in the 1860s (see Wakeling, *The Oxford Pamphlets, Leaflets, and Circulars of Charles Lutwidge Dodgson*).

[3] This article is by E. S. Leathes.

[4] This leaflet, first issued for Easter 1876, consists of only two short pages of text. It was reprinted on several occasions (and even in an unauthorized piracy) and was inserted into some, if not all, copies of the first edition of *The Hunting of the Snark*, which was also published in the spring of 1876.

[5] On 4 July 1862, Dodgson and his friend Robinson Duckworth took a boat trip up the

he finished converting it into *Alice's Adventures in Wonderland* (abbreviated in the nursery to *Alice in Wonderland*) in 1865; he published *Phantasmagoria*, which contained "Hiawatha's Photographing," in 1869; he finished *Through the Looking-Glass* in 1871, and *The Hunting of the Snark* in 1876. After that came a decline. His wit was as keen, his brain as masterfully intricate, as ever; but simplicity left him. Indeed, he never again quite caught the simplicity of his first book. *Alice in Wonderland* is an outpouring of inspired nonsense which flowed forth without hindrance and without perceptible impulse. But in *Through the Looking-Glass* we now and then hear the pump at work. The quality of the nonsense is no whit the worse; but simplicity is endangered. In *Through the Looking-Glass*, for example, there is the White Queen's exposition of living backwards, and the theory advanced by Tweedledum and Tweedledee that Alice and themselves had no existence apart from the Red King's dream—a perilous approach to metaphysics. Moreover, *Through the Looking-Glass* is a game of chess, which is the sheer superfluity of cleverness. But *Through the Looking-Glass* is only a shade less admirable than its companion. Has it not the White Knight and the two Queens, Tweedledum and Tweedledee, Humpty Dumpty, and the Walrus and the Carpenter? Has it not also the following passage, which has always seemed to us the perfect example of the higher foolishness?—

> 'Crawling at your feet,' said the Gnat . . . 'you may observe a bread-and-butter fly. Its wings are thin slices of bread and butter, its body is a crust, and its head is a lump of sugar.'
>
> 'And what does it live on?'
>
> 'Weak tea with cream in it.'
>
> A new difficulty came into Alice's head. 'Supposing it couldn't find any?' she suggested.
>
> 'Then it would die, of course.'
>
> 'But that must happen very often,' Alice remarked thoughtfully.
>
> 'It always happens.'

We may, indeed, feel quite certain of the longevity of the *Alice* books. They belong to no one period, but to all. They touch nothing actual but

---

Isis to Godstow with the Liddell children. In his diary he later wrote "On which occasion I told them the fairy-tale of *Alice's Adventures Underground*, which I undertook to write out for Alice."

human nature, and human nature is continuous and unchanging. Alice is a matter-of-fact, simple-minded child, and the world is full of Alices, and always will be. Hence the assured popularity of her history. Again, in the manner there is no sense of antiquity, although some thirty years have rolled by, each bringing its modification to literary style. Lewis Carroll wrote as plainly and luminously as he could; and we read and read and can think of no emendation whatever. The words are the best words in the best order. Of hardly any other humorist can it be said that in no instance do we ever wish his manner of narration altered. But Lewis Carroll was a merciless critic of himself and a tireless elaborator of his work, and he sent nothing forth until it was perfect.

By his art Wonderland is made not less conceivable than Fairy Land. It is almost impossible to believe that there is not somewhere such a region, where dwell for ever the Cheshire Cat and the Mock Turtle, the Gryphon and Humpty Dumpty, the Red Knight and the Duchess. They have each and all an individuality; and they are at once so mad and so reasonable: as real and recognisable as the people in Dickens. Partly it is Lewis Carroll's favourite trick of finding fun in pedantic literalness that persuades us. Again, the illusion is assisted by the abruptness with which the stories open. *Alice in Wonderland* has no preamble, there is no laboured description, we are in Wonderland in a moment, before there is time to think about the pinch of salt with which to season the exaggeration. These are the first words: "Alice was beginning to get very tired of sitting by her sister on the bank, and of having nothing to do," and then, on the third page, Alice has followed the white rabbit down the burrow. Again, in *Through the Looking Glass*, the beginning is immediate: "One thing was certain, that the white kitten had had nothing to do with it—it was the black kitten's fault entirely," and so on.

Alice in Wonderland has been translated into at least three European languages—French, German and Italian[1]—but without much success. Each country has its own humour and cares little for borrowing. In the title, at any rate, the German version bears the palm for conciseness: *Alice's Abenteuer im Wunderland.* The French and Italian are almost forbidding: *Aventures d'Alice au Pays des Merveilles* and *L'Avventure d'Alice nel Paese delle Meraviglie.* The two *Alice* books together were

[1] In fact, by the time of Dodgson's death, *Alice* had been translated into German (1869), French (1869), Swedish (1870), Italian (1872), Dutch (an 1875 abridged edition), Danish (1875), and Russian (1879). Of these only those mentioned in the article above were authorized by Dodgson. *Through the Looking-Glass* was not translated until after Dodgson's death, first into Italian ca. 1900.

converted to stage purposes some few years ago by Mr. Savile-Clarke, and the little play had an auspicious career both in London and the provinces.[1] Lewis Carroll took the keenest interest in this dramatic version—the stage, indeed, was among his hobbies—and when the company was at Brighton he journeyed thither and played fairy god-father (his favourite role in life) to some of the little performers. At that time a discussion was going forward in the papers concerning the proposed movement to make it illegal for children of less than ten years of age to appear on the stage, and Lewis Carroll, in a letter to the *St. James's Gazette*, referring especially to a meeting of ladies in favour of the movement, contributed to it.[2] The views of a man so fond of children and so passionately zealous for their happiness are peculiarly interesting. Here are extracts from his letter, which was entirely opposed to the projected measure:

> I spent yesterday afternoon at Brighton, where for five hours I enjoyed the society of three exceedingly happy and healthy little girls, aged twelve, ten, and seven. We paid three visits to the houses of friends; we spent a long time on the pier, where we . . . invested pennies in every mechanical device which invited such contributions and promised anything worth having, for body or mind, in return; we even made an excited raid on headquarters, like Shylock with three attendant Portias, to demand the 'pound of flesh'—in the form of a box of chocolate-drops—which a dyspeptic machine had refused to render. I think that anyone who could have seen the vigour of *life* in those three children—the intensity with which they enjoyed everything, great or small, that came in their way—who could have watched the younger two running races on the Pier, or have heard the fervent exclamation of the eldest at the end of the afternoon, 'We *have* enjoyed ourselves!'—would have agreed with me that here, at least, there was no excessive 'physical strain,' nor any *imminent* danger of 'fatal results'! . . . A drama, written by Mr. Savile-

---

[1] *Alice in Wonderland, a Dream Play for Children in two acts*, by Henry Savile Clarke (1841–93), opened at the Prince of Wales's Theatre on 23 December 1886 and ran through 18 March 1887. The production toured the provinces and was revived for the Christmas season in 1888 (see Lovett, *Alice on Stage*).

[2] Dodgson's letter, responding to an article which appeared on 16 July 1887, appeared in the *St. James's Gazette* on 19 July. Dodgson also published a letter on the subject in the *Sunday Times* on 4 August 1889 (reprinted in *The Theatre*, September 1889).

Clarke is now being played at Brighton; and in this (it is called 'Alice in Wonderland') all three children have been engaged. . . . They had been acting every night this week, and *twice* on the day before I met them, the second performance lasting till half-past ten at night, after which they got up at seven next morning to bathe! That such (apparently) severe work should co-exist with blooming health and buoyant spirits seems at first sight a paradox; but I appeal to anyone who has ever worked *con amore* at any subject whatever to support me in the assertion that, when you really love the subject you are working at, the 'physical strain' is absolutely *nil*; it is only when working 'against the grain' that any strain is felt; and I believe the apparent paradox is to be explained by the fact that a taste for *acting* is one of the strongest passions of human nature, that stage-children show it nearly from infancy, and that, instead of being, as these good ladies imagine, miserable drudges who ought to be celebrated in a new 'Cry of the Children,' they simply *rejoice* in their work, 'even as a giant rejoiceth to run his course.'"

From one who could write and believe:

> "Ah, happy he who owns that tenderest joy,
> The heart love of a child!"[1]—

these are striking words.

With *The Hunting of the Snark* (1876), which, although to most persons it seems more fitted to adult intellects, was dedicated by the author to a child, and frequently presented by him to children, Lewis Carroll's best period came to an end. Of this classic of comic verse it is hard to speak. No one has ever had a dream less coherent, less satisfying. Indeed, it may be said of Lewis Carroll that, above all men, he had the art of dreaming with a pen. His great colleague as a nonsense maker—Edward Lear[2]—could be foolish enough, but always with direction and with responsibility. Lewis Carroll, as does the mind when asleep, took

---

[1] From the dedication poem of *The Hunting of the Snark*, an acrostic on the name of Gertrude Chataway, one of Dodgson's child friends.

[2] (1812–88), English painter and author. Wrote *A Book of Nonsense* (1846) and popularized the Limerick. Though often mentioned as the twin pillars of 19th-century nonsense, Dodgson and Lear probably never met and were certainly not colleagues.

the line of least resistance. From *The Hunting of the Snark* illustrations have been excavated, by leader writers and politicians, for every kind of purpose; but the meaning of the complete work eludes us, and will elude. Because there is none.[1] It is simply fooling, the best fooling on record. Why, indeed, seek a meaning in a poem, when the preface to it can contain such a passage as this, in explanation of the line:

> "Then the bowsprit got mixed with the rudder sometimes"?

> The Bellman, who was almost morbidly sensitive about appearances, used to have the bowsprit unshipped once or twice a week to be revarnished, and it more than once happened, when the time came for replacing it, that no one on board could remember which end of the ship it belonged to. They knew it was not of the slightest use to appeal to the Bellman about it—he would only refer to his Naval Code, and read out in pathetic tones Admiralty Instructions which none of them had ever been able to understand—so it generally ended in its being fastened on, anyhow, across the rudder. The helmsman used to stand by with tears in his eyes; he knew it was all wrong, but, alas! Rule 42 of the Code, 'No one shall speak to the Man at the Helm,' had been completed by the Bellman himself with the words, 'and the Man at the Helm shall speak to no one.' So remonstrance was impossible, and no steering could be done till the next varnishing day. During these bewildering intervals the ship usually sailed backwards.

The resemblance in one of the illustrations to Dr. Kenealy, the Claimant's advocate, led some people at first to seek for a parable of the Tichborne Case.[2] Others have said that the Snark is popularity—"a boojum you see." But the story that the poem grew out of that line—

> "For the Snark was a boojum you see"—

---

[1] In response to one inquiry about the meaning of the *Snark*, Dodgson wrote, "I'm very much afraid I didn't mean anything but nonsense! Still, you know, words mean more than we mean to express when we use them."

[2] A long and famous trial involving a butcher from New South Wales who, in 1865, presented himself as Sir Roger Charles Tichborne, who had been lost at sea in 1854. The claimant to Tichborne's estate was eventually found guilty of perjury in 1874 (see Gardner, *The Annotated Snark*).

which one day flashed into the author's brain,[1] is the best explanation of all. In workmanship, *The Hunting of the Snark* is a miracle of dexterity.

After *The Hunting of the Snark* came a lull. Then there appeared, in 1883, *Rhyme? and Reason?* practically a reprint of *Phantasmagoria* and the *Snark*; *A Tangled Tale* (1885), a mixture of mathematical problems humorously enunciated, which were printed first in the *Monthly Packet*; *The Game of Logic* (1886), *Sylvie and Bruno* (1889), and, later, its second part, a whimsical medley comprising a story of modern life, a little exquisite nonsense—for example:

> He thought he saw a Banker's Clerk
>   Descending from the 'bus:
> He looked again, and found it was
>   A Hippopotamus.
> 'If this should stay to dine,' he said,
>   'There won't be much for us'—[2]

and much theology. *Sylvie and Bruno*, which grew from a little story contributed to *Aunt Judy* by Lewis Carroll in 1868,[3] was received with some disappointment, owing to the habit that readers have of demanding a favourite author to cut all his wares from the same piece. The theology was resented, not because it was not good—many of the passages are indeed beautiful and dictated by rare wisdom—but because it was considered to be out of place. Lewis Carroll, however, had grown to be of another opinion, and the two *Sylvie and Bruno* volumes were his favourites among his work.[4] In the same Easter greeting from which we have quoted at the head of this article he wrote (in 1876):

> I do not believe God means us to divide life into two halves—to wear a grave face on Sunday, and to think it out of place to even so much as mention Him on a week-day. Do you think He

---

[1] In his article "'Alice' on the Stage," in the April 1887 issue of *The Theatre*, Dodgson wrote "I was walking on a hill-side, alone, one bright summer day, when suddenly there came into my head one line of verse—one solitary line—'For the Snark *was* a Boojum, you see.'"

[2] From "The Mad Gardener's Song."

[3] The genesis of the *Sylvie and Bruno* books was the story "Bruno's Revenge" which appeared in *Aunt Judy's Magazine* (edited by Margaret Scott Gatty) in December 1867.

[4] The editors are not aware of any evidence to support this claim.

> cares to see only kneeling figures, and to hear only tones of prayer, and that He does not also love to see the lambs leaping in the sunlight, and to hear the merry voices of the children as they roll among the hay? Surely their innocent laughter is as sweet in His ears as the grandest anthem that ever rolled up from the 'dim religious light' of some solemn cathedral?

Lastly came, in 1896, the first part of *Symbolic Logic*, in which the young student is offered quite the most fascinating series of sorites ever propounded, where it is proved beyond all question, among other things, that "No Hedgehog takes in the *Times*."

Lewis Carroll has had many imitators—some quite shameless, and none worthy to stand beside him. They were, of course, doomed to failure, since they had neither his temperament nor his motive. Lewis Carroll, whose attitude to children was more devotion than mere affection, approaching even to adoration, was not a professional author: he was a kindly playmate of little people, and he wrote *Alice in Wonderland* to give pleasure to two friends, the little daughters of Dean Liddell, one of whom—the original Alice—is now Mrs. Hargreaves.[1] It was published that others might share that pleasure. Of not many of the diligent writers who have attempted to reap in the same field can it be said that their stories proceeded from a similar impulse. Indeed, the failure of the many imitations of *Alice* is another proof that good work must come from within, must be born of the author's own individuality. There has been, and can be, but one Lewis Carroll. To borrow his formulæ is not to reconstruct himself.

Lewis Carroll in private life was the Rev. Charles Lutwidge Dodgson, of whom we have hitherto said nothing, in accordance with his wish that his two characters should be kept apart. One proof of this desire is to be found in the letter which he wrote when, in 1888, Mr. R. H. Caine, the editor of a collection of humorous verse, asked him for permission to include certain of Lewis Carroll's pieces in that volume. Mr. Caine received this reply:

> Mr. C. L. Dodgson begs to say, in reply to Mr. Caine's letter received this morning, that he had never put his name to any such pieces as are named by Mr. Caine. His published writings

---

[1] Alice Pleasance Liddell married Reginald Hargreaves in 1880.

are exclusively mathematical, and would not be suitable for such a volume as Mr. Caine proposes to edit.[1]

Against this rebuff might be placed the following letter to a child (written in 1875) wherein the gulf existing between the two personalities is at once emphasised and removed; but it must be remembered that Mr. Dodgson would do for a child what he would not do for anyone else:

> My dear Magdalen,—I want to explain to you why I did not call yesterday. I was sorry to miss you, but you see I had so many conversations on the way. I tried to explain to the people in the street that I was going to see you, but they wouldn't listen; they said they were in a hurry, which was rude. At last I met a wheelbarrow that I thought would attend to me, but I couldn't make out what was in it. I saw some features at first. Then I looked through a telescope and found it was a countenance; then I looked through a microscope and found it was a face! I thought it was rather like me, so I fetched a large looking-glass to make sure, and then to my great joy I found it was Me. We shook hands, and were just beginning to talk when Myself came up and joined us, and we had quite a pleasant conversation. I said, 'Do you remember when we all met at Sandown?' And Myself said, 'It was very jolly there; there was a child called Magdalen,' and Me said, 'I used to like her a little. Not much, you know—only a little.' Then it was time for us to go to the train—and who do you think came to the station to see us off? You would never guess so I must tell you. They were two very dear friends of mine, who happened to be here just now, and beg to be allowed to sign this letter as your affectionate friends, Lewis Carroll and C. L. Dodgson.[2]

Mr. Dodgson was born in 1833, the son of a well-known Churchman, Archdeacon Dodgson.[3] He proceeded to Christ Church, Oxford, and in

---

[1] William Ralph Hall Caine not only printed this anecdote in the introduction to his 1890 volume *Humorous Poems of the Century*, but reprinted two of Dodgson's poems as well.

[2] The recipient of this letter was Magdalen Millard (1867–1921) whose father was a Fellow of Magdalen College, Oxford. Dodgson met both father and daughter in Sandown, Isle of Wight, on 5 October 1875, and wrote this letter on 15 December 1875, three days after visiting the family for tea (see Cohen, *The Letters of Lewis Carroll*).

[3] Charles Dodgson (1800–68), Student of Christ Church, Oxford, 1818–28, vicar at

1854 graduated with a first class in mathematics. In 1861 he was elected Senior student of his college, and in the same year became Mathematical Lecturer, a post he held until 1881. In 1861 he also took orders. His mathematical works were numerous and valuable, although his championship of Euclid against more modern systems of geometry is held by many to be fantastic. Mr. Dodgson had many of the eccentricities which so often accompany proficiency in his particular science, and many good stories are told of him at Oxford. He was a very watchful guardian of Oxford's honour, and used occasionally to put forth a whimsical pamphlet, in which some phase of the university's well-being was examined. These productions were always witty and marvellously ingenious. Mr. Dodgson was shy and reserved, a resolute celibate, a man of few friends but fit, and the patron saint of children. Incidentally we might mention that he liked them all to be familiar with Lewis Carroll's writings. His hobbies, after mathematics, which he looked upon both as work and play, were photography[1] and the stage. His photographs of children must be well-nigh countless. Mr. Dodgson—as sage, as wit, and as saint—will be mourned by those that knew him, as Lewis Carroll will be mourned by readers all the world over.

The Academy (London)[2]
January 22, 1898

My earliest sight of "Lewis Carroll" was when, as a freshman, raw and abashed, I had once the honour of sitting opposite him at dinner. With all a boy's nervousness at dining for the first time at a college "high table," in utter ignorance of the allusions which filled the

---

Daresbury, Rector of Croft from 1843, Canon of Ripon Cathedral from 1852, Archdeacon of Richmond from 1854, died 21 June 1868.

[1] Dodgson took up photography in the 1850s when the art was still in its infancy. Using the collodion wet plate process, he took thousands of photographs, including many portraits of children and famous adults. He also photographed little girls in "their favorite dress of nothing" on occasion, but only when both parents and subject consented. He gave up photography in the early 1880s when he chose to devote more time to writing.

[2] This article was signed "J. B."

talk, and tortured by a desire to escape to more congenial society, I found huge consolation in the fact that now I was regarding with my own eyes a god of my childhood. To one fresh from a very different place, and not yet habituated to the real Oxford, he seemed the living embodiment of the old Oxford of a boy's fancy. I desired to attend his lectures till I found that he was a mathematician. Dreary people in his own college, when questioned concerning their great man, confessed to having lived in ignorance that a prophet was among them. To certain he was simply an old mathematical tutor; to others a great name in letters which they had never connected with a local habitation; but to none was his figure noticeable. Few of Oxford's famous men have been so unconspicuous in her midst. Froude[1] was constantly to be observed; even Walter Pater[2] was known by sight to a large part of the undergraduate world; but I scarcely remember to have seen "Lewis Carroll" half a dozen times in the street.

In a sense he was the most old-world of all the elements in the place. The Oxford of ecclesiastical bustle and honest doubt, of Newman[3] and Mark Pattison,[4] of Arnold[5] and Clough,[6] though actually earlier in time, was years later in sentiment. And what shall I say of all that fills the gap between—the days of the new Liberalism, the aesthetic craze, the University Extensionist, the times which have "learnt a stormy note of men contention-tost, of men who groan," and are given over to many new things? The world of "Lewis Carroll" was ages removed from this. Though full of the wide human nature which delights in all things contemporary, his mind, alike in its piety, its ingenuities and its humours,

---

[1] James Anthony Froude (1818–94), Fellow of Exeter College, Oxford, 1842–49, editor of *Fraser's Magazine*, historian and author of *History of England from the Fall of Wolsey to the Death of Elizabeth*, also author of the anti-religious novel *The Nemesis of Faith*.

[2] (1839–94), Fellow of Brasenose College, Oxford, and the author of several books of essays and fiction, including *Studies in the History of the Renaissance* (1873) and *Marius the Epicurean* (1885).

[3] John Henry Newman (1801–90), Fellow of Oriel College, Oxford. Among the most influential theologians of the time, Newman was a leader of the Oxford Movement and in the 1840s left the Church of England and was ordained in the Catholic Church.

[4] Fellow of Lincoln College, Oxford, 1839–60, curator of the Bodleian Library, Rector of Lincoln College from 1861 until his death in 1884, and author of *Life of Casaubon*.

[5] Matthew Arnold (1822–88), Fellow of Oriel College, Oxford, Professor of Poetry, poet and literary critic, author of *Culture and Anarchy* and many other works.

[6] Arthur Hugh Clough (1819–61), Fellow of Oriel College, Oxford, poet, author of *Amours de Voyage* and other volumes.

belonged to an earlier and quieter world. His flute never lost "its happy country tone." His Oxford was sleepy and early Victorian, a haunt of people who played croquet and little girls with short frocks and smoothly brushed hair and quaint formal politeness. It seems to me that the exact subtlety of the humour of the *Alice* books could never be caught again, for the sleepy afternoon air, the quaint grace and the mock dignity are all the property of an elder and vanishing world.

In Oxford his works enjoyed a surprising popularity, and formed the storehouse for undergraduate nicknames. In my own day it even became the fashion for a man to set them in foolish paradox by the side of Shakespeare when incautiously questioned on his preferences in letters. *The Hunting of the Snark* was popularly supposed to contain all the metaphysics in the world. I once heard a distinguished college authority explain his course of lighter reading during one vacation. "The first week," he said, "I read *Sylvie and Bruno*." "And then?" some one asked. "And then," he murmured in doubt—"then," brightening up, "ah, then, I went back to *Through the Looking-Glass*."

The Adelaide Observer
January 22, 1898

The death is announced of the Rev. Charles Dodgson ("Lewis Carroll") the celebrated author, of Christ Church, Oxford.

[Mr. Dodgson has in recent years dropped into comparative obscurity, but at one time his nom de plume was familiar as a household word. The following were among his publications:—*A Syllabus of Plane Algebraical Geometry*, 1860; *The Formulæ of Plane Trigonometry*, 1861; *A Guide to the Mathematical Student in Reading, Reviewing, and Writing Examples*, 1864; *Alice's Adventures in Wonderland*, 1865; *An Elementary Treatise on Determinants*, 1867; *Phantasmagoria and other Poems*, 1869; *Songs from 'Alice's Adventures in Wonderland,'* 1870; *Through the Looking-glass and What Alice Found There*, 1871; *Facts, Figures, and Fancies Relating to the Elections to the Hebdomadal Council*, 1871; *Euclid, Book V., Proved Algebraically*, 1874; *The Hunting of the Snark: An Agony in Eight Fits*, 1876; *Euclid and his Modern Rivals*, 1879; *Doublets: A Word Puzzle*, 1879; *Rhyme? and Reason?* 1883; *A Tangled Tale*, 1885; *Alice's Adventures Underground*, 1886; *The Game of Logic*, 1887; *Curiosa Mathematica: Part I.—A New Theory of Parallels*, 1888; *Symbolic Logic*, 1896.]

Alliance News
May 6, 1898

The late Rev. Charles Lutwidge Dodgson, better known as "Lewis Carroll," the delightful author of *Alice in Wonderland*, died at The Chestnuts, Guildford, the residence of his sisters, in his 66th year. He was educated at Christ Church, Oxford, and distinguished himself in the schools, taking high mathematical and other honours. In 1861 he was ordained. After writing several mathematical works, *The Adventures of Alice in Wonderland* burst upon an astonished world. Few would have imagined that the quiet, reserved mathematician, a bachelor, who all his life was remarkable for his shyness and dislike of publicity, possessed the qualities necessary to produce a work which has stood the test of more than 30 years, and still captivates young and old alike by its quaint and original genius. *Alice in Wonderland* was originally written to amuse one of Dean Liddell's daughters. The author was an intimate friend of the Dean and Mrs. Liddell, and took infinite pleasure in the society of their little girls. It was in order to beguile her hours of playtime that these diverting fancies were woven for one of the children. The success of the book was never in doubt, and the story is current, though we cannot vouch for its authenticity, that the Queen herself, on reading it, was so much delighted that she commanded the author to send his next work to Windsor. He did so, and Her Majesty was almost as bewildered as Alice on finding that it consisted of *An Elementary Treatise on Determinants*![1] In many a home and many a schoolroom there was genuine sorrow when it was announced that the author of *Alice in Wonderland* and *Through the Looking Glass* had passed away.

The Athenaeum (London)
January 22, 1898

The late "Lewis Carroll" leaves behind him a gap in literature that will not soon be filled, for it may safely be said that of the many imitations evoked by the popularity of *Alice in Wonderland* not one has

[1] This anecdote was apparently repeated so often that Dodgson finally decided to refute it in print. In an advertisement first published in the second edition of *Symbolic Logic Part I* (1896), he wrote of the anecdote, "It is so constantly repeated, and is such an absolute fiction, that I think it worth while to state, once and for all, that it is utterly false in every particular: nothing even resembling it has ever occurred."

been a real success, nor has any one shown the power of acquiring that peculiar vein of humour which Mr. Dodgson possessed. It was too peculiar and personal, and was, in fact, largely connected with the mathematical cast of his intellect; nor, indeed, did he exhibit his singular combination of humour and fancy to much advantage except in *Alice in Wonderland* and *Through the Looking-Glass*. Even the *Hunting of the Snark* is not equal to its predecessors; and when he came to write *Sylvie and Bruno* theological matters dominated his mind, and caused the introduction of several tedious passages.

As a mathematician "Lewis Carroll" was quaint and original. His studies in logic were one of the chief pleasures of his recluse life, and when we reviewed his last publication (*Athen.* No. 3599), the first part of his *Symbolic Logic*, he was keenly interested, and addressed to us a long letter full of questions regarding our remarks. His views had at once the merits and defects that might be expected from him—thoughtfulness and ingenuity, mixed with a vein of caprice that rather impaired the merit of his speculations.

None of the newspapers seems to have noticed the origin of the pseudonym he used. "Lewis Carroll" is an anagram on his Christian names, Latinized Carolus Ludovicus.[1]

The Book World (London)
February 1898

The death on the 14th ult., at Guildford, in his sixty-sixth year, of the Rev. Charles Lutwidge Dodgson, mathematician, logician, theologian, and Oxford Don, but better known as "Lewis Carroll," the author of *Alice in Wonderland*, will doubtless be felt by a very large circle of both young and old as a real personal loss. He was also known as the writer of *Through the Looking Glass*—a sequel to the former-named book—*The Hunting of the Snark, Rhyme and Reason*, and *Sylvie and Bruno. Alice in Wonderland* was written to amuse one of Dean Liddell's daughters. As is abundantly evident from the long list of his works, "Lewis Carroll" was a serious and hard-working mathematician,

---

[1] In 1856, Dodgson sent a list of possible pseudonyms to Edmund Yates, editor of *The Train*, a comic magazine to which Dodgson contributed several pieces. Two of the possibilities were anagrams of his name (Edgar Cuthwellis and Edgar U. C. Westhill) and two were Latinized and re-Anglicized versions of Charles Lutwidge (Louis Carroll and Lewis Carroll). Yates chose Lewis Carroll.

and his chief title to fame will rest on those "Jeux d'Esprit" which have won for him a place in the affections of readers.[1]

The Bookman (London)
February, 1898

We understand that Lewis Carroll has left several books behind him ready for publication. It is, however, very doubtful whether they will all appear.[2]

The Bookman (London)
February, 1898

The death of "Lewis Carroll" has been followed by considerable enquiry for his charming books for children.

The Bookman, A Literary Journal (New York)
March, 1898

The Rev. Charles Lutwidge Dodgson, known as Lewis Carroll, the famous author of *Alice in Wonderland*, died on January 14th, at the home of his sisters, The Chestnuts, Guildford, in England, at the age of sixty-five. Mr. Dodgson was born at Croft, near Darlington, in 1833. His father was Archdeacon of Richmond. Mr. Dodgson had a fairly distinguished career at Oxford, especially in mathematics, and became a student of Christ-Church. He published various books on Mathematics and Logic, none of them very important, but his real fame began

[1] This final sentence is reprinted from *The Times* (15 Jan. 1898).

[2] *The Three Sunsets and Other Poems* had been prepared for the press by Dodgson before his death and was issued in February 1898 with illustrations by E. Gertrude Thomson. Most of the contents had been previously published. Dodgson left much material for the projected continuation of his *Symbolic Logic*. *Part I, Elementary*, had been published in 1896, but *Part II, Advanced* and *Part III, Transcendental* were not published in Dodgson's lifetime. William Warren Bartley III published *Lewis Carroll's Symbolic Logic* in 1977 and included a reconstruction of *Part II*. Other mathematical manuscripts remain unpublished and many papers of unknown content were burned after Dodgson's death.

with the publication of *Alice's Adventures in Wonderland* in 1865. This book was written for the daughters of Dean Liddell, and especially for Miss Alice Liddell, who died young.[1] It is a book of distinct genius, and stands by itself. It was followed by several others of unequal merit. Mr. Dodgson at one time was accustomed to give much of his time to entertaining children in his beautiful rooms, taking them to the play, and in other ways amusing them. But his life became sadder as the solitary years closed about him, and in later life he saw very little of children,[2] and greatly developed the eccentricities which had always to some extent characterised him. He practised the absurd affectation of refusing to be recognised as the author of his famous book.[3] He acted with studied discourtesy toward journalists, was peculiarly eccentric, and we fancy very troublesome in his way of publishing.[4] From many of his old friends he became entirely alienated,[5] yet for the good work which he did for children his name will long be remembered and honoured.

The following incident related of Lewis Carroll was characteristic of him, and bears out what we have just said about his eccentric behaviour. Last October a temerarious journalist called on Mr. Dodgson, and was received by him with a stately courtesy in his large study lined with books. He had evidently been reading in front of a blazing fire. His figure was

---

[1] In fact, Alice lived to the age of 82, dying in 1934, two years after she traveled to America to take part in the celebrations there of the centenary of Dodgson's birth. This article probably refers to the death of Alice's sister Edith Liddell shortly after her engagement to Aubrey Harcourt in 1876. Edith died from peritonitis at the age of 22 and is memorialized in a stained glass window in Christ Church Cathedral designed by Edward Burne-Jones (1833–98).

[2] In fact, though Dodgson ceased to accept official social invitations around 1880 in order to devote more time to his writing, he lead an active social life up to his death, frequently attending the theatre with both children and adults and inviting "child friends" (who in later years were often in their twenties—see below p. 68) to visit him for extended stays at Eastbourne where he passed his summers.

[3] Dodgson issued a circular letter (the "stranger circular") in 1890 which he used to disavow any connection with his famous pseudonym, writing that "Mr. Dodgson . . . neither claims nor acknowledges any connection with any pseudonym." While he went to lengths to avoid being publicly identified with "Lewis Carroll" and seemed to detest being lionized, he often used his famous identity as an entrée into the company of children and adults alike.

[4] This may be a reference to Dodgson's insistence, in 1883, that booksellers not be allowed to take such a large percentage of the profits on his books. Dodgson's books were financed out of his own pocket, and he felt that, taking the risk he should be due most of the reward. This led to an exchange of letters among booksellers in *The Bookseller*, some referring to "Lewis Carroll" as a dictator and others in kinder terms.

[5] The editors are aware of no evidence to support this claim.

slender and erect, his demeanour grave and reserved, and although somewhat worn, he seemed full of life. After a few words had been exchanged the visitor explained the object of his call, remarking that people would be greatly interested in reading a chat about his books, "especially *Alice in Wonderland*." Immediately at the utterance of that phrase Mr. Dodgson's manner completely changed, and he seemed to tremble with suppressed feeling. Rising from his seat and stuttering painfully, he said, "I cannot consent to anything of the nature of an interview. I do not acknowledge any book that has not my name upon it."

Lewis Carroll was a visitor at Tudor House[1] during Dante Gabriel Rossetti's tenancy, and in the new life of Christina Rossetti,[2] by Mr. Mackenzie Bell, we find the following extract from Miss Rossetti:

> With such inhabitants, Tudor House and its grounds became a sort of wonderland; and once the author of *Wonderland* photographed us in the garden. It was our aim to appear in the full family group of five; but while various others succeeded, that particular negative was spoilt by a shower, and I possess a solitary print taken from it in which we appear as if splashed by ink.

One of the photographs which "succeeded," as mentioned in this extract, is reproduced in Mr. Bell's book, and through the kindness of the publishers, Messrs. Roberts Brothers, we are permitted to present a facsimile of it to our readers.[3]

Boston Evening Transcript
January 15, 1898

Rev. Charles Lutwidge Dodgson, clergyman, mathematician and fairy-tale writer, is dead in London. Mr. Dodgson wrote two of the finest fairy tales of the world, as new and original as if nobody had ever written anything in that line before. An unalloyed gratitude belongs

---

[1] The house at 16 Cheyne Walk which was the London home of the poet and artist Dante Gabriel Rossetti (1828–92) from 1862 until his death in 1882.

[2] (1830–94) sister of Dante Gabriel and the author of *Goblin Market*. Christina Rossetti and Dodgson were admirers of one another's writings.

[3] The photograph printed with this article was one of Dodgson himself.

to this serious clergyman, who amid his mathematical work conceived these delightful adventures.

Mr. Dodgson was born in 1832, and had thus arrived at his sixty-sixth year. He was a first-class student in mathematics at Christ Church College, Oxford, in 1854; and was mathematical lecturer at Christ Church from 1855 to 1881. His works in his specialty are numerous and authoritative, and some of them are abstruse. It was in 1865 that he began as a romancer in fairy tales, with his immortal *Alice*, which Tenniel illustrated so charmingly. Afterward, in 1871, came the second *Alice* book, *Through the Looking Glass*, which was in no respect inferior to the *Adventures in Wonderland*. Then in 1876 "Lewis Carroll," as he called himself in this guise, issued *The Hunting of the Snark: an Agony in Eight Fits. Doublets: a Word Puzzle*; *Rhyme? and Reason?* a collection of verses mainly fantastic, came out in 1883; *A Tangled Tale* in 1885; *Silvie and Bruno* in 1888. The *Alice* books have gone through many editions, and so has the *Snark*, but of many of the lesser works one edition has sufficed, because he filled them with curious mathematical puzzles, as in *A Tangled Tale*, which was written for a periodical[1] and involved a contest for the solution of the puzzle contained therein.

Boston Evening Transcript
January 15, 1898

Rev. Charles L. Dodgson of Oxford, England, whose death is announced today, made his pen name of Lewis Carroll famous by one book, *Alice in Wonderland*, a modern English classic. On its face a book for children, it is in its inner significance a satire whose delicacy is most readily appreciated by grown people who have seen with their own eyes that dolls are stuffed with sawdust. The children can enjoy this famous book all the more from their ignorance of the world, while their elders from the opposite reason can return again and again to the perusal of the wonderful experiences that followed Alice's pursuit of the rabbit. Mr. Dodgson wrote other things in the same vein, and contributed to our language the "snark" and the "boojum." He also wrote sermons which we suppose will be read because their author was the creator of *Alice in Wonderland*.[2]

---

[1] *A Tangled Tale* appeared in *The Monthly Packet* from April 1880–May 1885.

[2] Only one sermon for children was published during Dodgson's lifetime (see below p. 47n); none of his other sermons are preserved.

One of the most profound mathematicians in Europe,[1] he gave a flat contradiction to Napoleon's dictum that mathematical gifts are inconsistent with the highest development of the imaginative faculty in their possessor. The story is told that Queen Victoria was so delighted with *Alice* that on ascertaining the name of the author, she ordered her bookseller to supply her with all Professor Dodgson's[2] works as soon as they came from the press, and received in her next book parcel a most erudite volume on the value of determinants. Dodgson was a good deal of a recluse in his private life, affording an illustration of the truth of the old saying that melancholy is a quality essential in the composition of the humorist.

The Brisbane Courier
January 17, 1898

The death is announced of "Lewis Carroll" (Mr. C. L. Dodgson), the well-known author.

The Chapbook (Chicago)
February 1, 1898

The Death of Lewis Carroll should not be passed over in silence. The Rev. Charles Lutwidge Dodgson, to give him his proper name, was born in 1832, graduated from Christ Church, Oxford, in 1854, with a first class in mathematics, and settled down in the university for the next twenty-six years as a mathematical lecturer. In 1881 he resigned and spent the remainder of his life in the charming village of Guildford in Surrey, doing some mildly clerical work.[3] He wrote and published

---

[1] Surely this is an exaggeration. Dodgson was not generally considered an important mathematician in his own time. For a thorough discussion of Dodgson's mathematical works, see Abeles, *The Mathematical Pamphlets of Charles Lutwidge Dodgson and Related Pieces.*

[2] Dodgson was a mathematical lecturer at Christ Church, but never held the post of university Professor. Still, he is incorrectly referred to as "Professor" in some obituaries.

[3] Though Dodgson resigned his mathematical lectureship in 1881, Christ Church remained his primary residence until his death. He did preach on rare occasions at St. Mary's Guildford (see above, p. 5n), but it would be inaccurate to describe him as doing "mildly clerical work." Guildford, even in the 1880s, was certainly a town, not a village.

several mathematical treatises of great value, but it is of course by his nonsense verses that he will be best remembered. *Alice in Wonderland* appeared in 1865, and was followed seven years later by its continuation, *Through the Looking Glass*. Then came the *Hunting of the Snark*—to mention only his best known works—and after that, in 1883, *Rhyme? and Reason?* Of these the two first were easily the most popular. They were translated into every European language, and reached all over the English-speaking world. Lewis Carroll had a command over rhythm as complete and a turn for phraseology as happy as either Calverley[1] or Praed;[2] but his chief virtue lay in his unique sympathy with childhood, and his power of rendering it with all its simplicity, incoherence, extravagance, and imaginativeness.

Christian Observer (Louisville, Kentucky)
February 2, 1898

Rev. C. L. Dodgson, better known under his nom de plume of Lewis Carroll, the author of *Alice in Wonderland*, died at Guildford, England, on Jan. 14. He was about seventy-five years of age. In 1865 he published that story of *Alice's Adventures in Wonderland*, which it is claimed has been read wherever the English language is spoken. It has been described as a "humorous fairy tale," but it is said to be a keen satire on the English politics of that day. Children enjoy it, but it takes the maturer child, or grown up intellect, to appreciate its wit and delicious nonsense. *Through a Looking Glass*, a continuation of Alice's adventures, and the *Hunting of the Snark*, were among his other notable books of clever nonsense. Mr. Dodgson was also the author of several mathematical text books.

Cleveland Press
January 15, 1898

A telegram from London announces the death of Professor Chas. L. Dodgson, of Oxford, Friday. Professor Dodgson was a clergyman and mathematician, born in 1832.

---

[1] Charles Stuart Calverley (1831–84). Poet and author of *Verses and Translations* (1862) and *Fly Leaves* (1872).

[2] Winthrop Mackworth Praed (1802–39). Poet and author of "The Vicar."

Although he never acknowledged the authorship, the popular children's stories of *Alice's Adventures in Wonderland*, *Davy and the Goblin*[1] and *Alice Through the Looking-Glass* were written by him, and have since become almost classics in their line.

His first book was published when he was 33 years of age,[2] and is enjoyed fully as much by older folks as by the children. One of his last books was a work on philosophy.[3]

Professor Dodgson was known as Lewis Carroll. The Dodo originated in his brain.[4] It is a marvelous creature.[5]

The Critic (New York)[6]
January 22, 1898

When, the other day, we were all reading the much-discussed list of *The Academy's* Academy, there were probably some, of a disposition supposed to be gravely serious, who were inclined to cavil at the inclusion of "Lewis Carroll's" name. It would have been quite useless to argue with such people. "You don't know much," the Duchess once said to Alice, "and that's a fact." Such would have had to be one's line of comment, whether they liked it or not—and it is on record that Alice did not much like the tone of the remark. One's only real feeling of incongruity about the appearance of Mr. Dodgson's name in the distinguished company comes from the fact that the *Alice* books have been such an institution for years, so much on a level with trial by jury and the Greek syntax, that it did seem to come with something of a shock to realize that their creator was still going about among men, even at the age of sixty-four.

---

[1] *Davy and the Goblin or What Followed Reading "Alice's Adventures in Wonderland,"* was written by Charles E. Carryl and published in the late 1880s.

[2] This clearly refers to *Alice's Adventures in Wonderland.* Dodgson had published several pamphlets and one book (*A Syllabus of Plane Algebraical Geometry,* 1860) prior to 1865.

[3] Possibly this refers to *Symbolic Logic* (a book on logic, not philosophy). Dodgson never published a book on philosophy.

[4] The Dodo was a real creature, native to the island of Mauritius before its extinction (by 1681). Dodgson and the Liddell children would have often seen the rare Dodo remains at the University Museum in Oxford and the 1651 painting of a Dodo by John Savery.

[5] This obituary was accompanied by a copy of the Tenniel illustration of Alice and the Dodo.

[6] This obituary was signed "A. I. Du Pont Coleman."

After all, though, it did not so much matter what one realized about this genial humorist, who passed away last week. The books were the thing, the substantive fact; and that *Alice* should have run by this time into the hundred thousand[1] is of more importance to the future of the race than that we should have known much about the kindly old Senior Student of Christ Church,

"Whose thoughts were full of indices and surds."[2]

He was an Oxford man through and through; the masterpiece itself is said to have been composed for the delectation of Dean Liddell's little daughters: and there are awestruck whispers which hint that certain very august personages in the University sat as unconscious models for some of its characters.[3] Many of the author's unpublished sayings are current there, deliciously humorous of course, but perhaps of too subtly academic a flavor to bear general reproduction.

Leaving aside *The Hunting of the Snark*, which is on one sustained level of excellence, and not venturing to say anything of his serious mathematical works, reputed to be educationally valuable, it is, after all, by the two *Alice* books that he will be forever known. *Sylvie and Bruno* opens bravely. The mob shouting for less bread and more taxes, the Professor's ingenious provision for a daily plunge-bath, seem to show that the author was wrong when he wrote that it would be courting disaster for him to attempt that style again;[4] but the "new path," of something suspiciously like "purpose-fiction," through which a good part of the book takes us, leaves us uncomfortably cold.

But *Alice*! And we do not mean the little girl herself, about whom the whole phantasmagoria revolves. Humpty Dumpty saw her weakness when he complained, "I shouldn't know you again if we did meet—you're so exactly like other people;" and as Stevenson[5] wrote of *The Vicomte de Bragelonne*,[6] "I may be said to have passed the best years of my life in

---

[1] At the time of Dodgson's death, the standard British edition of *Alice* had sold over 86 thousand copies and the People's edition close to 75 thousand. The comparable editions of *Looking-Glass* had sold over 61 thousand and nearly 50 thousand, respectively.

[2] From Dodgson's poem "A Riddle."

[3] Many attempts have been made to associate the characters in the *Alice* books with Oxford personalities and others. See, for examples, Gardner, *The Annotated Alice*.

[4] In the preface to *Sylvie and Bruno*, Dodgson wrote of the many imitators of *Alice*, "The path that I timidly explored . . . is now a beaten high-road . . . and it would be courting disaster for me to attempt that style again."

[5] Robert Louis Stevenson (1850–94). Scottish novelist and critic and author of *Kidnapped* and *Treasure Island*.

[6] 1850 novel by Alexander Dumas, père (1802–70). The second of two sequels to *The Three Musketeers*.

these six volumes, and my acquaintance with Raoul has never gone beyond a bow," so most of us, perhaps, merely tolerate Alice for the sake of the company into which she introduces us. Of that we shall never grow weary; and name after name of those friends is so dear and familiar among well brought-up people as to need no specific mention here. There is a whole philosophy of life within the old dark red covers.[1] If a motto were chosen for "Wonderland," nothing could be so apt as Oxenstiern's[2] "Videbis, fili mi, quantula sapientia mundus regitur";[3] and there is perhaps no book, after the two which are recorded to have helped Mr. Max Beerbohm[4]—Shakespeare and the Bible,—which supplies so many fitting quotations for every circumstance of our complex modern life.[5] And, though Lewis Carroll has "left off dreaming about" them, there is no fear that his beloved creations will "go out—bang!—just like a candle."

> Children yet, the tale to hear,
> Eager eye and willing ear,
> Lovingly shall nestle near,[6]

and we larger children, tired of finding men and women "so exactly like other people," shall find unceasing refreshment in the untrammeled fancy, in the kindly humor and the keen insight of a book which has already taken its place among the classics.

The Critic (New York) January 22, 1898

The following were published under the name of Lewis Carroll:—*Alice in Wonderland*, 1865; *Phantasmagoria*, 1869; *Through the*

[1] The standard editions of the *Alice* books issued during Dodgson's lifetime were bound in red cloth. In September 1864, while the book was being prepared for the press, Dodgson wrote to his publisher Alexander Macmillan, "I have been considering the question of the *colour* of *Alice's Adventures* and have come to the conclusion that *bright red* will be . . . the most attractive to childish eyes."

[2] Axel Gustafsson Oxenstierna (1583–1654), chancellor of Sweden.

[3] "You shall see, my son, with how little wisdom the world is governed."

[4] (1872–1956), English caricaturist, critic, and essayist. By 1898 he had written a book of essays *The Works of Max Beerbohm.*

[5] The *Alice* books are often cited as the most frequently quoted from in all the English language after Shakespeare and the Bible, though on what basis this claim is made or how such things can be counted is never mentioned.

[6] From the closing poem of *Looking-Glass*, an acrostic on Alice Pleasance Liddell.

Looking-Glass, 1872; *The Hunting of the Snark*, 1876; *Doublets*, 1879; *An Easter Greeting*, 1880; *Rhyme and Reason*, 1883; *Christmas Greetings* 1884; *A Tangled Tale*, 1885; *Alice's Adventures Under Ground*, being a reproduction of the manuscript book from which *Alice's Adventures in Wonderland* was developed, 1886; *The Game of Logic*, 1887; *Sylvie and Bruno*, 1889; *The Nursery Alice*, 1890; *Sylvie and Bruno, Concluded*, 1893; *Symbolic Logic*, 1896. The following books were published under the author's real name, Charles Lutwidge Dodgson: *An Elementary Treatise on Determinants*, 1867; *The Formulæ of Plane Trigonometry, Euclid and His Modern Rivals*, 1879; *Euclid, Books I and II*, 1882; *Lawn-Tennis Tournaments*, 1883; *Curiosa Mathematica*, 1888. It is related that when *Alice* was published, Queen Victoria read it, was much pleased with it, ordered that the author's next work should be brought to her, and the response to this command was *An Elementary Treatise on Determinants*!

We clip the appended reference to Mr. Dodgson from an article on Oxford celebrities contributed to *Harper's Monthly*, some years since, by Miss Ethel M. Arnold:[1]—

> It is perhaps, not generally known that Oxford is the home of 'Lewis Carroll,' the author of *Alice in Wonderland*, the queen of nonsense books. He is a senior student of Christ Church, and was for many years mathematical lecturer to the college, but retired from this latter post some few years ago, in order to devote himself more unreservedly to literary work. As might be gathered from his books, he is a genuine lover of children, and his beautiful suite of rooms in the northwest corner of Wolsey's great quadrangle, looking over St. Aldgate,[2] were at one time a veritable children's paradise. Never did rooms contain so many cupboards, and never did cupboards contain such endless stores of fascinating things. Musical boxes, mechanical performing bears, picture-books, innumerable toys of every description, came forth in bewildering abundance before the child's astonished eyes; no wonder, then, that in childish years a day spent with 'Lewis

[1] Ethel Margaret Arnold was one of Dodgson's child friends as was her sister Julia. The article appeared in the July 1890 issue of *Harper's Magazine* under the title "Social Life at Oxford," and was the source for the description of Dodgson's rooms in many of the obituaries printed here.

[2] Actually St. Aldates, one of the major thoroughfares of Oxford. This misspelling is repeated in several obituaries.

Carroll' was like a glimpse into a veritable El Dorado of innocent delights! For many years he was a considerable amateur photographer, and amused himself by taking his little friends in all sorts of odd and fanciful costumes, till his albums became filled with Japanese boys and girls, beggar-maids in picturesque tatters, or Joans of Arc in glittering armor. The smell of the collodion he used to pour on to the negative, his small 'subjects' watching him open-mouthed the while, lingers in the memory still, and the sight of the box in the dark room which used to be pulled out for them to stand upon, in order that they might watch more comfortably the mysterious process of 'developing,' served not long ago to remind one at least of his quondam child friends, humorously if a little painfully, of the flight of time.

Daily Graphic **January 20, 1898**

We greatly regret to announce the death, at Guildford, Surrey, in his sixty-sixth year, of the Rev. Charles L. Dodgson, known to the world by his nom de guerre of "Lewis Carroll."

Charles Lutwidge Dodgson was born in 1833, got his education at Christ Church, Oxford, graduated B.A., taking a first class in the Mathematical Schools, in 1854, became a senior student of Christ Church shortly afterwards, and took orders in 1861. His career furnishes almost the only, if not the only, instance known since the beginning of the world of an eminent mathematician who was also an eminent humourist. From 1855 to 1881—a period of six-and-twenty years—Mr. Dodgson was mathematical lecturer at Christ Church. During that period, and afterwards, he published several mathematical works, some of which were really valuable contributions to the literature of the higher branches of the subject. Among these are included: *An Elementary Treatise on Determinals* [*sic*], 1867, *Euclid Book V. proved Algebraically*, 1874, and *Mathematica Curiosa*, 1888. But he did not take even mathematics seriously, as witness his famous collection of comic postulates, of which the one that clings most tenaciously to the memory is to the effect that it must be granted that "A man can argue on any point at any distance from that point;"[1] and it was not long before he discovered that he was des-

[1] Probably meant to be a quote from Dodgson's humorous pamphlet *The Dynamics of*

tined to win fame, not as a reasoner but as a jester. In 1865 *Alice's Adventures in Wonderland* took the world by storm. Everybody was talking of "Tweedledum and Tweedledee,"[1] and quoting the famous parody of "You are old, Father William"; and the book was translated into almost every European language. *Through the Looking-glass and What Alice found there*, 1871, was somewhat less successful. It was a sequel, and there is generally something unsatisfactory about sequels; but it was a volume that would have made the reputation of any other humourist than "Lewis Carroll." The next work of importance was the famous *Hunting of the Snark*, 1876, which has continued for more than twenty years to be a joy to those who delight in all well-conceived absurdities. *Alice's Adventures Underground*, 1886, left the author's reputation where it was.[2]

Of all modern humourists Lewis Carroll was perhaps the most modest and retiring and the most tenacious of his pseudonym. Of his consistency in this regard a curious instance was furnished a few years ago. Mr. Ralph Caine, in preparing for a London publisher a book of elegant extracts of modern humorous poetry, was naturally anxious to include something from the pen of Lewis Carroll. So he wrote to the Rev. Charles Dodgson, asking permission to do so. In reply he received a type-written letter,[3] in the third person, explaining that the only works to which the Rev. Charles Dodgson had put his name were of a mathematical character, and entirely unsuitable for use in a publication of the kind which Mr. Caine contemplated. To anticipate the obvious remarks of the reviewers on the notable omission the editor told the story in his preface.

Lewis Carroll was never married, but, like many bachelors, he was very popular with children and very fond of them. Nothing pleased him better than to give children's parties in his Christ Church chambers. His intellectual energy continued unabated to the end. His last book, *Symbolic Logic*, was published as recently as 1896.

---

a Particle (1865). Under the heading "Postulates," Dodgson writes, "That a controversy may be raised about any question, and at any distance from that question." See Wakeling, *The Oxford Pamphlets, Leaflets, and Circulars of Charles Lutwidge Dodgson.*

[1] These characters do not appear in *Alice's Adventures in Wonderland*, but in *Through the Looking-Glass*.

[2] The preceding passage, beginning with "But he did not take" is copied verbatim from the January 15 article in *The Echo* (see below pp. 51–2).

[3] Dodgson did own and use a Hammond typewriter, which he acquired in 1888, but this is probably a reference to the "stranger circular" (see above p. 21n), a preprinted circular letter.

Daily Graphic
January 20, 1898

It may be an open question whether "Lewis Carroll," whose death we announce with deep regret, was the greatest humourist of his time; but there can be little doubt that he is the humourist whose loss the world will most deplore. The reason is, of course, the entire absence from his humour of that irony or bitterness which humourists too frequently affect. "Topsy-turvydom" was his note, as it is the note of his great contemporary, Mr. W. S. Gilbert;[1] but his worst enemy could never suggest that his "topsy-turvydom" sprang out of a lack of sympathy with persons or institutions which other people viewed with reverence. He joked, so to say, in the air, and at nobody's expense, and made the world laugh prodigiously at what was, after all, nothing but the vagary of a wayward fancy. The world will probably agree that the man who was able to do that had a higher genius than any of the humourists whose humour had its origin in satire.

The Daily Mail and Empire (Toronto, Canada)
January 17, 1898

By the death of Rev. C. H. Dodgson, known in literature as Lewis Carroll, one of the most sparkling of writers has been removed. Who has not read his *Alice in Wonderland* or the sister book *In a Looking-glass*? The unfortunate who has not yet laughed at the adventures of Alice ought to lose no time in joining in the merriment which, throughout the English-speaking world, the author of *Alice* has produced. A classic of the humorous description, *Alice*, whether in its prose or its poetry, is quaintness itself. The production of a clergyman given to serious studies, its quiet jocularity is all the more remarkable. Mr. Dodgson, it seems was as queer in his manners as in his writings. The cable tells us he never wore an overcoat.[2]

---

[1] William Schwenk Gilbert (1835–1911). Poet, dramatist, and critic, best known for his collaboration on a series of comic operettas with Arthur Sullivan.

[2] The cable went to many American and Canadian newspapers (see below, p. 95n).

The Daily News (London) January 15, 1898

In many a schoolroom, and in many a nursery, there will be sorrow for the death of "Lewis Carroll." Mr. Dodgson, to give him his real name, who died yesterday at Guildford in his sixty-sixth year, was with the possible exceptions of Thackeray[1] and Hans Andersen, the most successful writer of stories for children that the world has ever seen. *Alice in Wonderland* and *Through the Looking Glass* are books of which a child with an active mind never tires. They are equally full of imagination and of humour. They suggest so much more than they say that those who have grown up with them have found more in them every year. *The Hunting of the Snark* is not so universally popular. It is not quite such plain sailing. The reader has to do more for himself. He has, in the cant phrase, to read between the lines. But those who appreciate it, those who can taste it, as they need to say in the eighteenth century, love it with a tenderness that only increases with years. It may be doubted whether the favourite books of children are those which they thoroughly understand. They like the sense of mystery, of something beyond, something which they will be able to penetrate by-and-by. They certainly dislike the suspicion that a book has been written down to them, and that suspicion Lewis Carroll never allowed them to entertain. Moreover, he flattered the vanity of parents and guardians, of governesses and tutors. Most people, if they were candid, would acknowledge that they rather enjoyed the feeling that when they read to children they comprehended more than their audience. *Through The Looking Glass* is, among other things, a game of chess. But, apart from that technical and rather tiresome fact, Alice encounters problems which have perplexed the wisest in all ages. "How am I to get in?" she asked, when she came to the house of the white queen,[2] "Are you to get in at all?" said the footman, "That's the first question, you know." The footman was an Aristotelian philosopher without knowing it. He brought poor Alice to her bearings, and recalled her to first principles. There must be many of our readers who knew that the whiting blacked boots at the bottom of the sea long before it dawned upon them that there was a reason for entrusting that particular fish with that particular duty. "Reeling, and writhing, and fainting in coils," are such natural accomplishments for fishes that they hardly suggest at first

---

[1] William Makepeace Thackeray (1811–63), novelist and author of the popular children's story *The Rose and the Ring*.

[2] Actually, the house is that of the Duchess in *Alice in Wonderland*.

sight the corresponding accomplishments of dwellers upon the solid earth. The famous trial of the Knave of Hearts suggested only yesterday to the ingenious correspondent of a contemporary that the case against poor Dreyfus had been anticipated in that immortal burlesque.[1] Who that has sat listless and unoccupied among the junior bar in an Assize Court on a hot summer's afternoon can fail to recall the scene when he reads about the indictment of the pig on a charge of deserting his sty?

> The witness proved from the state of the straw
> That the sty was deserted when found,
> And the judge kept explaining the state of the law
> In a soft undercurrent of sound.

Mr. Dodgson was a man of genius. He was also a many-sided man. The son of an unusually theological Archdeacon, he was a strictly orthodox clergyman of the Church of England, and in politics a rigid Conservative. Most of his life was spent at Oxford, where he took honours in Mathematics, and became, as he remained, a tutor of Christchurch. He was a mathematician and the author of several mathematical works. There is indeed a story, perhaps too good to be true, that the Queen, being duly impressed with the excellence of *Alice in Wonderland*, desired Mr. Dodgson to send her a copy of the next book he wrote. Mr. Dodgson faithfully complied, as a loyal subject, with the gracious command of his Sovereign. As luck would have it, however, his next book was not *Through the Looking Glass*, but *An Elementary Treatise on Determinants*, which dealt with matters beyond even the wide range of her Majesty's interests and capacity. Mr. Dodgson was, among other things, a doughty champion of Euclid, whom he piously and successfully defended against the claims of modern rivals. He also took an active part in the rather provincial politics of Oxford, always as the enemy of innovation.[2] But his academic pamphlets, perhaps not much to his own annoyance, were read for amusement, and not for the sake of the side which they supported.

---

[1] Alfred Dreyfus (1859–1935), was an accused French spy whose court-martial and subsequent trials dragged on for years and created political crisis in France. In January of 1898, Emile Zola took up Dreyfus' cause in the press. For the complete passage as quoted in *The Times*, see below, pp. 174n, 175.

[2] While Dodgson did oppose some innovations, it is unfair to characterize him as the "enemy of innovation." Certainly on some topics he was very much in favor of innovation. In matters of College and University business, he judged issues individually, rather than taking a "standard" stance.

When he invited old and young Oxford to find the value of an examiner in terms when there was no examination, or contrasted the man who took in twelve books and got a third with the man who took in the examiners and got a second,[1] the controversy in aid of which these puzzles were invoked was extinguished in laughter. Hundreds of Oxford men must remember the weird appearance of "the House" when the Cathedral was undergoing what architects call restoration. Many of them will recollect also "The Three T's, or the Trench, the Tunnel, and the Tea-Chest,"[2] with which Mr. Dodgson celebrated the occasion. The Tea-Chest was the temporary belfry—"style early debased, very early and extremely debased."[3] The innocent stranger walking towards the Cathedral by the Oberwell through Christchurch meadows saw "a number of right angles, and a number of anglers who had no right to be there."[4] There had been a recent performance of Bach's Passion Music in the sacred edifice, for which so many tickets had been issued that late comers were unable to find admission. "Call you this Baching of your friends?"[5] asked the indignant tutor. Oxford never cared what Mr. Dodgson's opinions might be. Everyone bought his pamphlets, sure of a hearty laugh at fun which never sprang from malice, and scarcely ever gave pain. Much of the humour was verbal. The creatures behind the looking glass and in wonderland used to drive Alice mad by verbal repartee. "I can't have any more tea," said Alice, "because I've had none at all." "You mean you can't have any less," replied the March Hare. But in all humour except the highest, the verbal element is large.

The two *Alices* were, as is well known, written for the daughters of Dr. Liddell, the accomplished, distinguighed, and imposing personage who was Dean of Christchurch for so many years. Alice herself has long been dead. Dr. Liddell survives a relic of old classical Oxford at the height of its glory. For many of the brief and fleeting generations which

---

[1] Paraphrases from *The Dynamics of a Particle* (1865). See Wakeling, *The Oxford Pamphlets, Leaflets and Circulars of Charles Lutwidge Dodgson.*

[2] Dodgson's pamphlet *The Vision of the Three T's*, which lampooned the architectural changes in Tom Quad, was published in 1873.

[3] From *The New Belfry of Christ Church, Oxford* (1872).

[4] A misquotation of *The New Belfry*. The quote actually reads "There the thoughtful passer-by, with four right angles on one side of him, and four anglers, who have no right to be there, on the other. . ." Christ Church meadow is bordered on one side by the river Isis, hence the reference to anglers.

[5] Epigram of *The Vision of the Three T's* (1873), the line is a parody of Shakespeare, *King Henry IV, Pt. 1*, act II, scene iv, "call you that backing of your friends?"

renew the youth of a University, Mr. Dodgson was a familiar figure in the streets of Oxford. He looked the very type of a prim, austere, clerical Don. But if any children caught sight of him, they broke from all restraint and dashed at him madly. He had no children and no wife of his own. But to children of all sorts and sizes he was an irresistible magnet. He attracted them partly, of course, because he petted them and made much of them, partly because he told them stories and amused them, but partly also, and perhaps mainly, because he entered into their minds and hearts. The gift of childhood, whatever it be, remained with him like the peace of God, with which it is perhaps identical, always. He never saw it "die away, and fade into the light of common day."[1] Child worship is the essential quality of his best books. If it is sometimes impossible not to laugh at Alice it is never possible not to love her. She is, unlike most children in story-books, a mere child, without any mixture of womanhood or girlhood, without any ideas which are not absolutly childish. Yet she fascinates people of all ages provided only that they are not destitute of imagination and humour. If they are so unfortunate, they pronounce the books an "insult to their intelligence" and fling them down in a rage. Since the *Hunting of the Snark*, which appeared more than twenty years ago, Mr. Dodgson has produced nothing worthy of it or its predecessors. *A Tangled Tale* is forgotten. The two volumes of *Sylvie and Bruno* are little read. His verse, except for the poems introductory to *Alice* and *The Looking Glass*, was not of high order, and *Phantasmagoria*, despite its clever parody of Longfellow,[2] is a dead book. But the first two *Alices*, the only *Alices* worth mentioning, keep, as they deserve to keep, their old place. They have never been superseded. It is possible that they never will be. There are no "topical allusions" in them, and children do not alter, even if men and women do, which may be doubted. In private life, and with grown up people, Mr. Dodgson was taciturn. His inference from the Homeric controversy that "the Iliad and Odyssey were not really written by Homer, but by another person of the same name," has become almost proverbial. But it is as the children's friend that he will live, and it may be said of him on the morrow of his death, that he "never wrote a line which dying, he would wish to blot."

---

[1] William Wordsworth (1770–1850), *Ode—Intimations of Immortality from Recollections of Early Childhood* (1807), line 76.

[2] "Hiawatha's Photographing" is a parody of "The Song of Hiawatha" by Henry Wadsworth Longfellow (1807–82).

Daily Telegraph (London)[1]
January 15, 1898

Yesterday, at The Chestnuts, Guildford, the residence of his sisters, peacefully passed away the Rev. Charles Lutwidge Dodgson, better known as "Lewis Carroll," and better still as the author of *Alice in Wonderland*, *Through the Looking-Glass*, and *The Hunting of the Snark*. As a writer, "Lewis Carroll" belonged to no officially recognised category; he broke entirely fresh ground—was at once a humourist, a poet, a satirist, a mathematician, a polished man of the world, and a clergyman. In a word, like the German, Jean Paul Friedrich Richter,[2] he richly deserved the epithet of the unique. If ever wit and humour were born with a man it was with Mr. Dodgson, whose clever repartees, sudden sallies, and comical comparisons took the form of "winged words," and circulated from Land's End to John o' Groats. But it is mainly as a writer for children that he will live in English literature. In *Alice in Wonderland* and *Through the Looking-Glass* "Lewis Carroll" roofed the children's fantastic world with wonder, fun, and fairy grace. The voices he raised in that Kingdom of Nowhere will not die so long as the language of Shakespeare continues to be spoken. The nursery has no more welcome guests than those two volumes, which "enrich the threshold of the long wintry night with showered largess of delight," and render the little ones unconscious of the sliding hour. They are eminently tales which hold the children from play and old men from the sleep of the chimney corner.

Mr. Dodgson, who was born in 1833, entered Oxford University in 1854, and became a senior student of Christ Church. No such curious combination of the humourist and mathematician had ever before resided within the precincts of that or any other University. His knowledge of higher mathematics, which was extensive and profound, enabled him to discharge the duties of Mathematical Lecturer from 1855 to 1881, and at one time it was generally expected that he would be raised to the University chair of that science. He was the author of several bright and amusing mathematical pamphlets, one of which—*Euclid and the Game of Logic*[3]—well deserves to be republished. In one of those writings he

---

[1] Portions of this obituary appeared in the *St. James's Gazette* (see below pp. 131–5). Portions were also reprinted in the *Darlington and Stockton Times* (see below pp. 40–44).

[2] German novelist (1763–1825). Author of *Titan* (1800–1803).

[3] Apparently a mistaken portmanteau of Dodgson's two works, *Euclid and His Modern Rivals* (1879) and *The Game of Logic* (1886)

parodies the propositions of Euclid with inimitable wit. Take, for instance, the unanswerable, self-evident, and lamentable proposition: "A discussion may be raised on any point at any distance from that point." Or this other, which should be carefully borne in mind by all public orators of all countries at the present day: "A plain, superficial speaker is one who, any two points being given, will lie evenly between those points."[1]

The book which made his name famous, *Alice in Wonderland*, was written in two Long Vacations;[2] and the principal character, Alice, was believed to be modelled on one of the daughters of the Dean of Christ Church. The quiet quaintness and pedantic precision which characterise the wild whimsicalities of the dramatis personæ of this fairy land of nonsense are unrivalled, and the verses, "You are old, Father William," are as popular as the most finished productions of our classics. The story rapidly circulated throughout the English-speaking world, and was translated into all the languages of Europe. A large number of clever persons of the class who found or sought a key to Carlyle's[3] *Sartor Resartus* discovered that *Alice in Wonderland* was a far-fetched illustration of a game of chess![4] Seven years later (1872) appeared the continuation of *Alice* under the title *Through the Looking-Glass*, and although only a sequel, it was scarcely inferior to its predecessor, and was almost equally appreciated. These two works are masterpieces of the exuberant fancy, graceful style, and poetic genius of Lewis Carroll. Nothing that he wrote before or after—and numerous ingenious literary efforts have been since associated with his name—come at all near to these. The verses entitled "The Jabberwock,"[5] which are composed in a language as artificial as "Volapuk"[6] or "Esperanto,"[7] are none the less familiar to and understanded

---

[1] This misquotation from Dodgson's humorous pamphlet *The Dynamics of a Particle* (1865), completely robs the excerpt of its humor. The original reads, "Plain Superficiality is the character of a speech, in which any two points being taken, the speaker is found to lie wholly with regard to those two points." Unfortunately, this misquotation appears in many obituaries.

[2] Dodgson originally wrote the book as *Alice's Adventures Underground*, a manuscript which he began on 5 July 1862 and finished on 13 November 1862. The later published version of the story, *Alice's Adventures in Wonderland* was much expanded.

[3] Thomas Carlyle (1795–1881), Scottish critic and historian. His novel *Sartor Resartus* was published in 1833.

[4] *Looking-Glass*, not *Alice*, is based on a game of chess.

[5] "Jabberwocky," is the title of the poem which appears in *Looking-Glass*.

[6] An artificial language for international communication invented around 1879.

[7] Artificial language created by Ludwik Lazar Zamenhof and presented in 1887.

of the nursery. The "Walrus and the Carpenter," which also appeared in *Through the Looking-Glass*, is even more popular than "Old Father William."

Of all his subsequent works, among which we may mention the collection of parodies and poems entitled *Phantasmagoria* (1889), *Doublets*, *Euclid and his Modern Rivals*, *A Tangled Tale*, and the *Game of Logic*, the most widely appreciated was *The Hunting of the Snark.* His latest work, *Sylvie and Bruno,* must, on the contrary, have come as a disappointment to his friends. Whereas his earlier works transport children to a fairy realm of fun and frolic, in which not even the transient shadow of a moral flits across the perpetual sunshine of delight, a large portion of *Sylvie and Bruno* is taken up with what must be termed sermons and lessons, which utterly spoil whatever humour is interwoven with the story. The author of *Sylvie and Bruno* is not the Lewis Carroll who composed *Alice in Wonderland*, but the Rev. Charles Lutwidge Dodgson, clergyman.

Mr. Dodgson was to the last an uncommonly pleasing and popular man, whose conversation was often suggestive and always amusing. Despite the duties of his calling, which he discharged with scrupulous care, he had numerous acquaintances in the theatrical world of London, who respected his learning, humoured his amiable eccentricities, and keenly appreciated his wit. The sayings attributed to him at Oxford would fill an entertaining volume of Carrolliana. Among other things, his "etymology of the bell"[1] is still quoted with relish by scholars. There was a provisional belfry at Christ Church College, which was familiarly known to Oxonians of the time as "the meat safe." Mr. Dodgson, undertaking to explain this epithet etymologically, split up the word belfry into two parts—the French word belle and the German word frei (free). Then he went to work as follows:

Belle=beautiful=comely=meet (meat);
Frei=free=secure=safe
Result: "Meat-safe."

One of his best repartees was elicited by an Oxford student driving a tandem. This candidate for academical distinction had the social honour of driving the late Khedive of Egypt through the ancient City of Oxford, and the misfortune to have "spilled" his Highness. On the following day the youth, driving the same tandem, overtook Mr. Dodgson, who looked

[1] The etymology is taken from Dodgson's humorous pamphlet *The New Belfry of Christ Church* (1872) and was printed under the heading "On the etymological significance of the new Belfry, Ch. Ch."

somewhat tired. "May I give you a lift?" he asked, jauntily. To whom the humourist: "Wilt thou slay me as thou didst the Egyptian yesterday?"[1]

In truth, Lewis Carroll was moulded of the porcelain clay of human kind. His salient characteristic consisted in a curious combination of poet, mathematician, and man of the world, such as has probably never been seen before. The man, "Lewis Carroll" has passed away from among us, but his works are destined to endure.

The Daily Telegraph (Sydney, Australia) January 17, 1898

London, Saturday.—The death is announced to-day of the Rev. Charles Dodgson ("Lewis Carroll"), the author of *Alice in Wonderland* and a number of other books.

The late Rev. Charles Dodgson was senior student of Christ Church, Oxford. He became better known under the pseudonym "Lewis Carroll," and endeared himself to juvenile hearts as the author of the delightful *Alice's Adventures in Wonderland*, *Through the Looking-glass*, and several other books of a similar character. He also devoted his attention to graver subjects, and published volumes bearing such sober titles as *A Syllabus of Plane Algebraical Geometry*, *The Formulae of Plane Trigonometry*, *Symbolic Logic*, etc., the last-mentioned book having been published in 1896.

Darlington and Stockton Times[2] January 22, 1898

The death is announced of the Rev. Chas. Lutwidge Dodgson, better known as "Lewis Carroll," and better still as the author of *Alice in Wonderland* and other works.

It is not generally known that this famous writer is a Croft man. He was the son of the Rev. Mr. Dodgson, who held the living of Croft, near

---

[1] This anecdote, much repeated in the press, was disclaimed by Dodgson's sister (see below p. 160). Dodgson's alleged reply is a paraphrase of Exodus 2:14, "Intendest thou to kill me, as thou killedst the Egyptian?"

[2] Much of this obituary appeared in the *Daily Telegraph* (15 Jan 1898). See above p. 37–40 for full footnotes.

Darlington, for some 20 years,[1] in succession to the Rev. Mr. Dalton,[2] and who was also Archdeacon of Richmond and Canon of Ripon. His second name he took from his mother, who was, prior to her marriage to the rector, a Miss Lutwidge.[3] It is about 30 years since Archdeacon Dodgson died, but several members of his family, including two daughters and the Rev. Skeffington Dodgson,[4] have always kept up a visiting connection with the district. His succesor at Croft was the Rev. Mr. Law,[5] who afterwards changed livings with the Rev. Mr. Villiers,[6] whose death took place a few years ago.

The author of *Alice in Wonderland* was born at Croft in 1833, and spent his early days at the Rectory. In 1854 he entered Oxford University and became a senior student of Christ Church. He was ordained deacon in 1861, but never took priest's orders. His reason for this was said to be a slight hesitancy of speech which prevented his speaking in public. This, however, he in a measure overcame, and he not infrequently read the lessons and prayers at the college services in Oxford Cathedral. He died on Friday, Jan. 14th, at the Chestnuts, Guildford, the residence of his sister.

No such curious combination of the humourist and mathematician had ever before resided within the precincts of that or any other University. His knowledge of higher mathematics, which was extensive and profound, enabled him to discharge the duties of Mathematical Lecturer from 1855 to 1881, and at one time it was generally expected that he would be raised to the University chair of that science. He was the author of several bright and amusing mathematical pamphlets, one of which—*Euclid and the Game of Logic*—well deserves to be republished. In one of those writings he parodies the propositions of Euclid with inimitable wit. Take, for instance, the unanswerable, self-evident, and lamentable proposition: "A discussion may be raised on any point at any distance from that point." Or this other, which should be carefully borne in mind by all public orators of all countries at the present time: "A plain, superficial speaker is one who, any two points being given, will lie evenly between those points."

---

[1] Rev. Dodgson held the living at Croft for 25 years, from 1843–68.

[2] James Dalton was Rector of Croft from 1805–43.

[3] Frances Jane Lutwidge.

[4] Skeffington Hume Dodgson (1826–1919), Charles Dodgson's brother.

[5] Frederick Henry Law, Rector of Croft 1868–73.

[6] Charles Villiers (born Lawrence), Rector of Croft 1873–94.

The book which made his name famous, *Alice in Wonderland*, was written in two Long Vacations; and the principal character, Alice, was believed to be modelled on one of the daughters of the Dean of Christ Church. The quiet quaintness and pedantic precision which characterise the wild whimsicalities of the *dramatis personæ* of this fairy land of nonsense are unrivalled, and the verses, "You are old, Father William," are as popular as the most finished productions of our classics. The story rapidly circulated throughout the English-speaking world, and was translated into all the languages of Europe. A large number of clever persons of the class who found or sought a key to Carlyle's *Sartor Resartus* discovered that *Alice in Wonderland* was a far-fetched illustration of a game of chess! Seven years later (1872) appeared the continuation of *Alice* under the title *Through the Looking-Glass*, and although only a sequel, it was scarcely inferior to its predecessor, and was almost equally appreciated. These two works are masterpieces of the exuberant fancy, grace style, and poetic genius of Lewis Carroll. Nothing that he wrote before or after—and numerous ingenious literary efforts have been since associated with his name—come at all near to these. The verses entitled "The Jabberwock," which are composed in a language as artificial as "Volapuk" or "Esperanto," are none the less familiar to and understanded of the nursery. The "Walrus and the Carpenter," which also appeared in *Through the Looking-Glass*, is even more popular than "Old Father William."

Of all his subsequent works, among which we may mention the collection of parodies and poems entitled *Phantasmagoria* (1889), *Doublets*, *Euclid and his Modern Rivals, A Tangled Tale*, and the *Game of Logic*, the most widely appreciated was *The Hunting of the Snark*.

As a writer, "Lewis Carroll" belonged to no officially recognised category; he broke entirely fresh ground—was at once a humourist, a poet, a satirist, a mathematician, a polished man of the world, and a clergyman. In a word, like the German, Jean Paul Friedrich Richter, he richly deserved the epithet of the unique. If ever wit and humour were born with a man it was with Mr. Dodgson, whose clever repartees, sudden sallies, and comical comparisons took the form of "winged words," and circulated from Land's End to John o' Groats. But it is mainly as a writer for children that he will live in English literature. In *Alice in Wonderland* and *Through the Looking-Glass* "Lewis Carroll" roofed the children's fantastic world with wonder, fun, and fairy grace. The voices he raised in that Kingdom of Nowhere will not die so long as the language of Shakespeare continues to be spoken. The nursery has no more welcome guests than those two volumes, which "enrich the threshold of the long wintry night with showered largess of delight," and render the little ones unconscious of the sliding hour. They are eminently tales which hold the children from play and old men from the sleep of the chimney corner.

Mr. Dodgson was to the last an uncommonly pleasing and popular man, whose conversation was often suggestive and always amusing. Despite the duties of his calling, which he discharged with scrupulous care, he had numerous acquaintances in the theatrical world of London, who respected his learning, humoured his amiable eccentricities, and keenly appreciated his wit. The sayings attributed to him at Oxford would fill an entertaining volume of Carrolliana.

Mr. Dodgson's beautiful suite of rooms in Christ Church College were a real wonderland for many a real child. From his innumerable cupboards musical boxes, mechanical performing bears, picture-books innumerable, toys of every description, came forth in bewildering abundance before the child's astonished eyes. Nor were Mr. Dodgson's little friends all of one class. He took for many years a deep interest in Homes and Hospitals for Sick Children, and no inconsiderable portion of the profits of his books was devoted to these works of charity.[1] He was to the last an uncommonly pleasing and popular man, whose conversation was often suggestive, and always amusing. Despite the duties of his calling as mathematical tutor at Christ Church, he had numerous acquaintances in the theatrical world of London, who respected his learning, humoured his amiable eccentricities, and keenly appreciated his wit.[2] Among other things, his "etymology of the bell" is still quoted with relish by scholars. There was a provisional belfry at Christ Church College, which was familiarly known to Oxonians of the time as "the meat safe." Mr. Dodgson, undertaking to explain this epithet etymologically, split up the word belfry into two parts—the French word belle and the German word frei (free). Then he went to work as follows:

Belle=beautiful=comely=meet (meat);

Frei=free=secure=safe

Result: "Meat-safe."

One of his best repartees was elicited by an Oxford student driving a tandem. The young gentleman had the social honor of driving the late Khedive of Egypt through the ancient city of Oxford, and the misfortune to have "spilled" his Highness. On the following day the youth, driving

---

[1] The profits from the facsimile of *Alice's Adventures Underground* were so donated, and Dodgson frequently made presents of his books (though sometimes of rejected editions considered unfit for sale) to such hospitals.

[2] This repetition of an earlier passage is indicative of the danger of pastiche obituary.

the same tandem, overtook Mr. Dodgson, who looked somewhat tired. "May I give you a lift?" he asked jauntily. To whom the don replied, "Wilt thou slay me, as thou didst the Egyptian yesterday?"

A story which may or may not be true is that the Queen, on reading *Alice in Wonderland*, was so much delighted that she commanded the author to send a copy of his next work to Windsor. He did so, and Her Majesty was almost as bewildered as Alice on finding that it consisted of *An Elementary Treatise on Determinants*!

The Dial (Chicago) February 1, 1898

There is such a thing as nonsense," says Walter Bagehot[1] in one of his essays, "and when a man has once attained to that deep conception, you may be sure of him ever after." This is no doubt a true saying, but it needs to be supplemented. To attain to the conception in question is only the beginning of wisdom, for there should go with the realization that nonsense exists a healthy instinct for its enjoyment when it is of the innocuous sort, and for its vigorous repudiation when it assumes a pernicious form. There is in most human nature a streak of irrationality which makes people an easy mark for certain forms of very dangerous nonsense—for the pseudo-scientific jargon of palmistry or theosophy, let us say—and hopeless indeed is his case who becomes entrapped by these pitfalls. Yet the irrational instincts that find such vent belong to the normal psychological make-up of mankind; they are survivals, probably, although a neo-Schopenhauerianism might claim for them a deeper cause than that, but in any case they have to be reckoned with. Most people chafe, consciously or unconsciously, under the limitations of a world rigidly subjected to the principle of casuality, and seek at times to escape into some world in which fancy may rove untrammeled, in which life may seem less strenuous than science makes it out to be.

This is a somewhat ponderous exordium for a few observations about the work of that gentle "maker" of nonsense whose recent death has made childhood of all ages the poorer wherever the English language is spoken, but it was peculiarly the gift of "Lewis Carroll" to open our eyes to

[1] (1826–77). English social scientist and editor of *The Economist* from 1860 until his death.

just such a "land of wonder-wander" as we all need to see at times, and need all the more in proportion as our habitual outlook is uncompromisingly logical and matter-of-fact. If we cannot take refuge in some such world of transparent impossibilities we are in danger of gratifying our deep-seated irrational instincts by straying off into delusions of the sort that paralyze the powers of thought, of confusing vagaries with realities, and of dissipating our intellectual substance in speculations that cannot possibly lead to anything; whereas we may give ourselves up to the charm of Alice and her Wonderland without the least compunction, and do not run the slightest danger of taking too seriously the doings of the Mad Hatter, the March Hare, or the Queen of Hearts. If these things are "to the Greeks foolishness," we are sorry for the Greeks, and remain unshaken in our loyalty to the imaginings of the chronicler of Alice, the poet of the Boojum and the Jabberwock.

> A little nonsense now and then
> Is relished by the wisest men.[1]

so runs the old saw, and to "Lewis Carroll," purveyor of the most delightful nonsense in our language, a debt of the deepest gratitude is due.

Of that eminent clergyman and distinguished mathematician, the Rev. Charles Lutwidge Dodgson, the public knows little beyond an occasional amusing anecdote, like that which relates how the Queen, charmed with the story of Alice, requested the author to send her his other works, and was not a little surprised when she received a collection of mathematical treatises. He was born about sixty-five years ago, took his degree at Christ Church, Oxford, and remained a college don for the greater part of his life. *Alice's Adventures in Wonderland* was published in 1865, *Through the Looking-Glass* in 1872, and *The Hunting of the Snark* in 1876. Among his other books that belong to literature, we may mention *Phantasmagoria*, *Doublets*, *Rhyme? and Reason?*, *Euclid and His Modern Rivals*, *A Tangled Tale,* and *Sylvie and Bruno*. The three books first mentioned have not only become classics of the nursery, but have also won a place among the books with which every cultivated reader is expected to have some acquaintance. Their characters and incidents are quoted "to point a moral and adorn a tale"[2] almost as frequently as the characters and inci-

---

[1] Old nursery rhyme of unknown authorship.
[2] Samuel Johnson (1709–84), author/lexicographer, *Vanity of Human Wishes* (1749).

dents that appear in the novels of Dickens or the poems of Tennyson.[1] They have added new words to the common English vocabulary, and people make use of "beamish" and "galumphing" and "chortling" as if familiar concepts. Such verses as "The Walrus and the Carpenter" and "How doth the little crocodile" are as well known as the best lyrics of the more serious poets; while the story of the slaying of the Jabberwock rivals in its haunting mystery the story of Childe Roland and the Dark Tower,[2] besides being rather more intelligible. It is no ordinary triumph to have thus imposed a new literary genre upon the consciousness of the reading public, and the extent of the achievement is best realized when we compare the success of "Lewis Carroll" with the comparative failure of even the best of his many imitators.

The books that are concerned with childhood may be divided into three classes. The first and largest of these classes includes the books that are written for children alone, that have no message at all for the adult reader. The second class includes those few books that state things from the childish point of view, but are really intended for the delight of children of a larger growth. *A Child's Garden of Verses*[3] and *The Golden Age*[4] are the two matchless examples of this kind of literature. Here we have the freshness of feeling that remembers what being a child was like, conjoined with an exquisite literary art which revives that feeling in the grown-up reader, who could not possibly summon it back into consciousness unless thus aided. But the child, although he may take some interest in these books, cannot possibly enjoy them as the "Olympians" do. The books of the third class, to which belong those which we owe to the creator of *Alice*, are equally appreciated by young and old (although of course in different ways), and make for a better understanding between the ages, because they create a new bond of sympathy, setting the child and the man for the time being upon a common ground of interest.

---

[1] Alfred, Lord Tennyson (1809–92), poet laureate and author of *Idylls of the King* and *In Memoriam.* Dodgson met Tennyson in 1857 in the Lake District and on several other occasions at the poet's home, Farringford, in Freshwater Bay, Isle of Wight.

[2] "Childe Roland to the Dark Tower Came," a poem by Robert Browning (1812–89), published in his *Men and Women* (1855).

[3] Book of poetry published in 1885 by Robert Louis Stevenson (1850–94).

[4] Book of short stories published in 1895 by Kenneth Grahame (1859–1932), author of *The Wind in the Willows* (1908).

Eastbourne Gazette[1]
January 19, 1898

The Rev. C. L. Dodgson for some years past has been a seat holder and a most regular worshipper at Christ Church during the two or three months he spent in Eastbourne, which was his favourite seaside resort. He told me that ours was the only church he really cared for in the place, and he rarely ever missed an attendance morning or evening. Of his work as a clergyman I know very little. Of late, however, I believe he officiated very rarely in church. Last summer, at my suggestion, he gave several addresses at our Children's Services, which were greatly appreciated, and when thanking me for the opportunity he said it was the first time that he had addressed an audience of children under similar circumstances. The subject he took was an original allegory, entitled "Victor and Arnion," its purpose being to illustrate the temptations to which children are exposed, and to teach them how they may be avoided and conquered. The addresses were delivered with deep feeling; at times the speaker was scarcely able to control his emotions, especially when speaking of the love and compassion of the Good Shepherd. I understood him to say that he intended to elaborate and ultimately to publish the allegory.[2] Besides these addresses I suggested that Mr. Dodgson should speak to the various departments of our day schools, which he consented to do with the proviso that he would try one as an experiment. This proved so satisfactory to himself as well as to the teachers and children, that he went the round, giving much pleasure to all. On these occasions he told a series of humorous and entertaining stories with illustrations on the blackboard. Mr. Dodgson, I believe, never classified himself with any particular school of thought in our Church, but his sympathies and convictions lay much more with the Evangelical section than any other. He was essentially an independent thinker and adopted his own lines in this

---

[1] This article, "The Late Lewis Carroll. A Great Humourist and a Devout Churchman. The Author of *Alice in Wonderland*. His Connection with Eastbourne," was signed "W. H. H." and written by Rev. William Henry Hewett, vicar of Christ Church, Eastbourne, beginning in 1890. Dodgson spent his summers in Eastbourne from 1877 until his death.

[2] A similar address, given at St. Mary Magdalen Church, St. Leonards-on-Sea, a few miles away on 3 October 1897, was printed in the *St. Mary Magdalen Church Magazine* for November 1897 and reprinted in Collingwood, *The Lewis Carroll Picture Book*. It was the only publication in his lifetime of a work by Dodgson which identified his real name with his pseudonym (which was spelled "Louis Carroll" in the church magazine).

as in other things. If I were asked from my own observation to mention two prominent characteristics connected with his religious life I should point to the depth of his reverence and the simplicity of his faith. One manifestation of the former was seen in his untiring protest against the modern system of gabbling through our church service, especially the General Confession. This he held, most appropriately, should be said with great deliberateness of utterance and humility of tone. Accordingly he would never allow himself to be hurried in its repetition, with the result that sometimes before he had finished his Confession the clergyman would have got some way through the Absolution. This, he said, was almost invariably his experience at Oxford. This method of holding his own against an unfortunate habit of too many of our clergy may be considered eccentric, and perhaps unsuitable, but all who share his views in this particular will allow that it was most reasonable. It was certainly genuine and I believe unostentatious. With regard to his faith none who heard the series of addresses he gave at Christ Church to the children would question the spiritual simplicity of his confidence in the Gospel revelation of a personal and ever present Saviour to be known in the life of man through the grace of submission faith, and obedience.

Eastbourne Gazette[1]
January 19, 1898

Within a few yards of "Great Tom,"[2] up the winding stairs above the Junior Common Room, and overlooking the great Quad of Christ Church, lived the late author of "Alice in Wonderland."

Many were the undergraduates who felt that they knew him well, though most of them perhaps had never spoken to him, as his well-known figure day by day might be seen in "Peckwater"[3] or "Tom Quad," or taking his solitary walks by the river or in the parks.

In the frost of a January morning, when the sun had risen only enough to tinge with rose and grey the winter twilight, or in November when the

---

[1] This article, "'Lewis Carroll' at Christ Church," was signed "An Appreciation by an Undergraduate."

[2] Great Tom is the bell which hangs in Christopher Wren's Tom Tower above the entrance to Christ Church, and which rings 101 times each night at 9:05 to commemorate the original students of The House.

[3] Peckwater Quad is the quadrangle in front of the library at Christ Church.

mists from off the meadows hung like a great sheet over all things, every morning, summer or winter, his familiar face was seen at Matins, sitting in his stall beyond the Dean, putting to shame many an undergraduate, aye, and Don too, who preferred the warmth of his cozy rooms to the unpleasant walk to chapel through the cold cloisters.

Yes, his place in Cathedral and "Hall" will be sadly missed next term by many of us who used to look for his kindly features and familiar figure.

Alice in Wonderland, that unique book which delights equally the child in the nursery and those who long since have lost their fairylands, was given to the world by accident. Many years ago it was Charles Dodgson's delight to take in his boat up the Cherwell, on summer afternoons, Dean Liddell's three children; and having moored her to the bank, beneath the shade of one of the many willows that overhang the stream, to recount to his little listeners the wonderful adventures of Alice. As his story grew, he wrote it down, illustrating it with his own hand. Long after, a friend, picking up the manuscript, persuaded the writer, much against his will, to publish it.[1] Thus it came before the public, with new illustrations by Tenniel, and considerably altered from its original form.

A service beginning at 8:15 p.m. on a Sunday night in October is not very attractive to the ordinary undergraduate, who at that hour looks forward to his pipe, coffee, and an arm chair. But the galleries of the University Church were well filled to hear Charles Dodgson the few times that he has preached of late years. Not many of those who had the pleasure of hearing him, will soon forget his sad and scholarly face, as he delivered his remarkable and unique address (the last, as it proved to be) on Eternal Punishment.[2] The hard thinking of many years was so evident in every sentence that he uttered in his musical and somewhat slow words.

Those who know Oxford well, will know that among the many members of its community the scouts are the most neglected. It was for these men that Charles Dodgson had a special service every Sunday, which he himself conducted.

And perhaps, besides the thousands of men, women, and children who have been cheered and charmed by his unique books, may be counted

---

[1] George MacDonald (1824–1905), fantasy writer and author of *At the Back of the North Wind*, was one friend who encouraged Dodgson to publish *Alice*.

[2] Dodgson's sermon on 24 October 1897 was on reverence (see below pp. 116n, 119). Dodgson did not believe in the doctrine of Eternal Punishment, and doubtless expressed his feelings on the subject if not in this sermon in some other one.

the many manservants whose lives have been brightened by his cheering words and scholarly counsels during term in Oxford.

It may be interesting to those who see in the coming age the great era of unrecognised Truths being made manifest to a materialistic generation to know that "Lewis Carroll" was a member of the Psychical Research Society and that the books which lined the walls of his bedroom were nearly all of them volumes relating to psychic phenomena in its many different forms.[1] His especial interest was the investigation of hypnotic and will power manifestations. He told me the last time I had the privilege of paying him a visit in his quiet sanctum.

So has passed away from us, at the age of 66, that two-fold personality: on the one hand, the recluse, Don, mystic, mathematician of the highest order, of all men the most reserved and reserving; on the other, the humorist, author, devoted lover of little children, poet, and Christian clergyman, who brought smiles and sunshine wherever he went.

The "House" has produced many great men since its foundation some hundred years ago,[2] but time alone will prove whether our Alma Mater can be capable of giving the world another genius like "Lewis Carroll."

Eastbourne Gazette[3]

The late "Lewis Carroll" (the Rev. C. L. Dodgson), the author of *Alice in Wonderland*, was a frequent visitor to Eastbourne, and I am told that he was in the habit of staying at a house in that central yet quiet and sequestered thoroughfare, Lushington-road.[4] I learn that he was a regular attendant at Christ Church, that he preached on at least one occasion at the Ocklynge Mission Room,[5] and that he was recently in

---

[1] Dodgson left thousands of books on a myriad of subjects at the time of his death (see Stern, *Lewis Carroll Bibliophile*). Which books in particular lined his bedroom walls cannot be known.

[2] Christ Church was founded as Cardinal's College in 1525, refounded as King Henry VIII's College in 1532, and founded as Christ Church, a dual foundation of college and cathedral, in 1546. Perhaps the phrase should have read "some hundreds of years ago."

[3] From an undated clipping in the Dodgson family album.

[4] Dodgson lodged at Number 7, Lushington Road, every year from 1877 through 1895. When his landlady Mrs. Dyer moved to Number 2, Bedford Well Road, in late 1895, Dodgson followed her and lodged there for his last two summers.

[5] On 10 September 1893, Dodgson wrote in his diary, "Preached my first Eastbourne sermon: at 'Ocklynge Mission-Room'. I took as text *Luke xi, 4*—'Lead us not into temptation'. The room seemed full: perhaps two hundred were there."

correspondence with the Vicar of Eastbourne. The delightful humour of Mr. Dodgson appeals alike to the young and old. He never wrote a line that could offend the most fastidious refinement; and, being a deeply religious man, he was guiltless of that habit of attempting to jest on serious subjects to which writers who are devoid of real wit are prone.

The Echo (London) January 15, 1898

We regret to announce the death of the Rev. Charles Lutwidge Dodgson, M.A., known to so many thousands of readers, old and young, as "Lewis Carroll." Mr. Dodgson was staying with his sisters at their residence at Guildford, where he died yesterday in his 66th year.

A Mathematician and an Author.

Mr. Dodgson and Lewis Carroll were in authorship, and to some extent in life also, two very different beings. Mr. Dodgson was the don, and writer on mathematical subjects; Lewis Carroll was the author of charming books for children. Mr. Dodgson graduated at Christ Church, Oxford, in 1854—having taken a first class in mathematics, and a third in Lit. Hum. In the following year he was appointed Mathematical Lecturer to the College, a post which he continued to occupy until 1881. In 1858 he was elected to

A Senior Studentship,

and this he continued to hold until the end. In 1861 he was ordained. His earliest, as well as his latest, publications, were all mathematical—*A Syllabus of Plane Algebraic Geometry* (1860*), The Formulæ of Plane Trigonometry* (1861), *A Guide to the Mathematical Student* (1864). The first of his children's books was published in 1865. Other mathematical works followed, of which the more important were—*An Elementary Treatise on Determinants* (1867*), Euclid, book V., proved Algebraically* (1874), and *Curiosa Mathematica: A New Theory of Parallels* (1888).

Fame as a Jester.

But he did not take even mathematics seriously, as witness his famous collection of comic postulates, of which the one that clings most tenaciously to the memory is to the effect that it must be granted that "A man can argue on any point at any distance from that point;" and it was not long before he discovered that he was destined to win fame, not as a reasoner but as a jester. In 1865 *Alice's Adventures in Wonderland* took the world by storm. Everybody was talking of "Tweedledum and

Tweedledee," and quoting the famous parody of "You are old, Father William;" and the book was translated into almost every European language. *Through the Looking-glass and What Alice found there*, 1871, was somewhat less successful. It was a sequel, and there is generally something unsatisfactory about sequels; but it was a volume that would have made the reputation of any other humourist than "Lewis Carroll." The next work of importance was the famous *Hunting of the Snark*, 1876, which has continued for more than twenty years to be a joy to those who delight in all well-conceived absurdities. *Alice's Adventures Underground*, 1886, left the author's reputation where it was.[1]

The Evening Post (New York) January 15, 1898

The Rev. Charles Lutwidge Dodgson, better known by his pen name of Lewis Carroll, the author of *Alice in Wonderland*, died yesterday in Guildford, England. He was born in 1832, was graduated from Christ Church College, Oxford, in 1854, with a first class in mathematics, took orders in 1861, and from 1855 to 1881 was lecturer in mathematics. He published *Alice's Adventures in Wonderland* in 1865, which, with its continuation *Through the Looking-Glass* (1872), became a nursery classic, and was translated into many tongues. He also published a collection of poems and parodies entitled *Phantasmagoria* (1869), *Hunting of the Snark* (1876), *Doublets* (1879), *Rhyme? and Reason?* (1883), *Euclid and His Modern Rivals* (1879), *A Tangled Tale* (1886), *Game of Logic* (1887), and *Mathematica Curiosa* (1888), the last being a valuable mathematical treatise.[2] He also wrote other mathematical works.

Eminent as were his abilities as a mathematician, and solid as were his achievements in scientific works it is by his nonsense verse that his name will be preserved to posterity. These had extraordinary qualities, quite independent of their wonderfully happy imitation of childish simplicity, incoherence, extravagance, and imaginativeness, such as a peculiarly happy phraseology, and a most uncommon command of rhythm. Their popularity extends all over the English-speaking world.

---

[1] This paragraph was reprinted in a piece in the *Daily Graphic* (20 January 1898). See notes above, pp. 30–31.

[2] The wording of this list of works is nearly identical to that in the *Evening Star* (Washington, D.C.) (see below p. 53).

The Evening Star (Washington, D.C.) January 15, 1898

A dispatch from London yesterday announces the death of the Rev. C. L. Dodgson, whose nom de plume was Lewis Carroll, the author of *Alice in Wonderland.*

Charles Lutwidge Dodgson (Lewis Carroll), an English clergyman, mathematician and writer, whose wonderful and fantastic verses for children have charmed the youngsters of the English speaking world, was born in 1832. He entered Christ Church College, Oxford, and was graduated in 1854 with a first class in mathematics. He was elected a student of his college, took orders in 1861 and from 1855 to 1881 was lecturer in mathematics.

Under the pen name "Lewis Carroll" he issued in 1865 *Alice's Adventures in Wonderland*, which, with its continuation, *Through the Looking Glass* (1872), became a nursery classic and was translated into many tongues.

He also published a collection of poems and parodies entitled *Phantasmagoria* (1869*), Hunting of the Snark* (1876), *Doublets* (1879), *Rhyme? and Reason?* (1883), *Euclid and His Modern Rivals* (1879), *A Tangled Tale* (1886), *Game of Logic* (1887), and *Mathematica Curiosa* (1888), the last being a valuable mathematical treatise.

The combination of mathematics and fantastic verse has long been regarded as curious and remarkable. In both he excelled, but it is as Lewis Carroll, and not as churchman or mathematician, that he will live longest.

The Gentlewoman[1] January 29, 1898

Stand still, true poet that you are!
I know you; let me try and draw you.—Browning.

In trying to sketch this outline of my beloved friend—one of the most unique and charming personalities of our time—my own shadow, so to speak, must perforce, to my regret, sometimes fall across

[1] This article, titled "Lewis Carroll. A Sketch by an Artist-Friend," was written by Emily Gertrude Thomson (1850–1929). Thomson met Dodgson, as described in this

the picture. I shall not attempt to give any of his incomparable humour, all the world knows that through his immortal books, but I want to show that little known view of him—that "other side of the moon"—visible only to those who knew him best.

It was at the end of December, 1878, that a letter, written in a singularly legible and rather boyish-looking hand, came to me from Christ Church, Oxford, and signed "C. L. Dodgson." The writer said that he had come across some fairy designs of mine, and he should like to see some more of my work. By the same post came a note from my London publisher (who had supplied my address) telling me that the "Rev. C. L. Dodgson" was "Lewis Carroll."

Alice in Wonderland had long been one of my pet books, and as one regards a favourite author as almost a personal friend, I felt less restraint than one usually feels in writing to a stranger, though I carefully concealed my knowledge of his identity as he had not chosen to reveal it.

This was the beginning of a frequent and delightful correspondence, and as I confessed to a great love for fairy lore of every description, he asked me if I would accept a child's fairy-tale book he had written called *Alice in Wonderland*. I replied that I knew it nearly all off by heart, but that I should greatly prize a copy given to me by himself. By return came *Alice*, and *Thro' the Looking Glass*, bound most luxuriously in white calf and gold.

And this is the graceful and kindly note that came with them:—

> I am now sending you 'Alice' and the 'Looking Glass' as well. There is an incompleteness about giving only one, and besides the one you bought was probably in red and would not match these. If you are at all in doubt as to what to do with the (now) superfluous copy, let me suggest your giving it to some poor sick child. I have been distributing copies to all the hospitals and convalescent homes I can hear of, where there are sick children capable of reading them, and though, of course, one takes some pleasure in the popularity of the books elsewhere, it

---

article, in 1879 and the two were close friends for the remainder of his life. Known for her drawings of fairies, Thomson studied at the Manchester School of Art and was a member of the Royal Society of Miniature Painters. Her illustrations adorn Dodgson's posthumously published collection of verse *Three Sunsets and Other Poems* (February, 1898). The article was accompanied by two photographs of Dodgson, one by W. Shawcross of Guildford (which was reproduced in several obituaries), and the other of Dodgson standing at a table. The epigram is from "Popularity" and refers to Keats.

is not nearly so pleasant a thought to me as that they may be a comfort and relief to children in hours of pain and weariness. Still, no recipient *can* be more appropriate than one who seems to have been in fairyland herself, and to have seen, like the 'weary mariners' of old—

> Between the green brink and the running foam
> White limbs unrobed in a crystal air,
> Sweet faces, rounded arms, and bosoms prest
> To little harps of gold.

"Do you ever come to London?" he asked in another letter; "if so, will you allow me to call upon you?"

Early in the summer I came up to study, and I sent him word that I was in town. One night coming into my room, after a long day spent at the British Museum, in the half-light I saw a card lying on the table. "Rev. C. L. Dodgson." Bitter, indeed, was my disappointment at having missed him, but just as I was laying it sadly down I spied a small T.O. in the corner. On the back I read that he couldn't get up to my rooms early or late enough to find me, so would I arrange to meet him at some museum or gallery the day but one following? I fixed on South Kensington Museum,[1] by the "Schliemann" collection,[2] at twelve o'clock.

A little before twelve I was at the rendezvous, and then the humour of the situation suddenly struck me, that *I* had not the ghost of an idea what *he* was like, nor would *he* have any better chance of discovering *me*! The room was fairly full of all sorts and conditions as usual, and I glanced at each masculine figure in turn, only to reject it as a possibility of the one I sought. Just as the big clock had clanged out twelve, I heard the high vivacious voices and laughter of children sounding down the corridor.

At that moment a gentleman entered, two little girls clinging to his hands, and as I caught sight of the tall, slim figure with the clean-shaven, delicate, refined face, I said to myself, "*That's* Lewis Carroll." He stood for a moment, head erect, glancing swiftly over the room, then bending down, whispered something to one of the children; she, after a moment's pause, pointed straight at me.

---

[1] Now the Victoria and Albert Museum.

[2] Collection of artifacts collected by the German archaeologist Heinrich Schliemann (1822–1890), best known for his excavations at ancient Troy and Mycenae.

Dropping their hands he came forward, and with that winning smile of his that utterly banished the oppressive sense of the Oxford don, said simply, "I am Mr. Dodgson; I was to meet you, I think?" To which I as frankly smiled and said, "How did you know me so soon?"

"My little friend found you. I told her I had come to meet a young lady who knew fairies, and she fixed on you at once. But, *I* knew you before she spoke."

Thus began that close and happy friendship which gladdened and enriched my life for eighteen years, and ended only when *his* beautiful life drew to its close, scarcely a week ago.

Soon after our meeting in London he wrote from Oxford: "Are you sufficiently unconventional (I *think* you are) to defy Mrs. Grundy, and come down to spend the day with me at Oxford? Write and ask permission of your father." Needless to say the permission was given, and the visit arranged.

"What would you like to eat?" he wrote. "Choose your own lunch, and, whether possible or impossible, it shall be got!" I went down by an early train, and he met me at the station with two more little girls. Always little girls! We did the lions in Oxford. Glorious, of course. Then the children were taken home, and we returned to his beautiful rooms at Christ Church. (To be concluded.)

The Gentlewoman
February 5, 1898

Lewis Carroll was not only an admirable amateur photographer, but an enthusiastic sketcher of children, especially when they were "dressed in nothing," as he called it, and, *apropos* of this, he once told me an amusing remark of one of Sir Noel's [*sic*] Paton's[1] children. They were very beautiful, and served their father as models in those two exquisite illustrations for Kingsley's *Water Babies*.[2] In the design of the fairies floating through the water the front view figure is an absolute portrait of one little daughter. One day a friend in looking at it said to the

---

[1] Sir Joseph Noël Paton (1821–1901). Painter especially known for fairy scenes and mythological and religious subjects. Dodgson approached him to illustrate *Looking-Glass* before convincing John Tenniel to undertake the work, but Paton declined.

[2] Charles Kingsley (1819–75). English author and clergyman. His *The Water Babies* (1863) was a popular children's book which just preceded *Alice* into print.

child, "Why, that's *you.*" "Yes," was the reply, "it's *me*, but I don't *often* dress like that!"

I consider that he naturally had a decided gift for drawing, but he was entirely untrained, so that his sketches, though they had a certain feeling for beauty, were, of course, very crude. He had a singularly correct eye for form, so much so that he would instantly detect slight flaws in drawing in some of the work I did for him, which had escaped my more practised eye. Complimenting him one day on his ability as art critic, he said, "I can't draw in the least myself—that's the first qualification for an Art Critic. One approaches a subject in such a delightfully open and unbiassed manner if you are entirely ignorant of it!"

Writing to me once on this subject he said: "I *love* the effort to draw; but I fail utterly to please even my own eye. But I have no time now left for such things. In the next life, I do hope that we shall not only *see* lovely forms, such as this world does not contain, but shall also be able to *draw* them."

He was always anxious that whatever work I did for him should be drawn from life. "I do want you to do my fairy-drawings from *life*," he wrote. "They would be very pretty, no doubt, done out of your own head, but they will be ten times more valuable if done from life. Mr. Tenniel is the only artist who has drawn for me who resolutely refused to use a model, and declared he no more needed one than *I* should need the multiplication table to work out a mathematical problem!"

Soon there followed another day at Oxford.

"Come and photograph human fairies," he wrote.

His photographic studio, on the roof of the college,[1] was a big place filled with all sorts of properties, costumes, &c. He dressed up the children in a variety of quaint costumes, and "took" them in all sorts of attitudes; intervals for refreshment and play being very frequent. The magic cupboards were opened and there issued forth a marvellous procession. Mechanical bears and wrestlers, rabbits, monkeys, and other uncouth and delightsome beasts. We would group together on the floor, Lewis Carroll, the fairies, the beasts and myself, and gay indeed were the hours we spent. How his laugh would ring out like a child's!

---

[1] Dodgson had obtained special permission to erect a "glass house" on the roof above his rooms in Tom Quad at Christ Church, Oxford. The wet-plate photography process which he used required as much light as possible, and his rooftop studio provided just that. Although he stopped photographing in the early 1880s, the studio remained, and was not removed until after his death (see above p. xiii).

And the exquisite nonsense he talked! It was like pages out of the *Alices*, only more delightful, for there was his own voice and smile to give the true charm to it all. I used to try to recall and record it. It was impossible—as impossible as to catch the gleam of colour on sunlit water, or to grasp a drifting rainbow. It was a mystic, intangible, gossamer-like thing, that, to chain it down in the words with which we should have translated it, would have been to crush out all life and grace—to destroy it altogether.

His love for children was not merely for the amusement they afforded him; it was deeper far than that. It was their purity and innocence that appealed to him; and if ever a child showed by word or look any unlovely feeling, the change in his face was startling; all the light would die out of it, and his tender reproof was inexpressibly touching.

To show his sensitiveness of conscience and dread of evil for children, I will quote a passage from one of his letters. I had a little girl model, about eight years of age, whose lovely head I had sketched and sent to him, and as I was at the time working at some fairy designs for him, in which all the little figures were to be "dressed in nothing," I wrote, "I mean to ask Edie to sit to me as a 'fairy'—that is, if she doesn't mind."

I must quote from memory, for the letter is mislaid. He wrote to say that if Edie showed the slightest shrinking from this, that he must implore me not to try to overcome it. If she had such a feeling, it would be a crime in the sight of God to persuade her to consent.

One day he came to my studio to sketch a very charming child whom I had engaged specially for him. She had travelled some little distance from the North of London to come to us, and I expressed some surprise that her mother had allowed her to come alone, but the child said she was "used to it."

On his return to Oxford he wrote to me:—

"I don't quite like the idea of that small and pretty child going all that way alone on my account. If she got lost or stolen I should feel an awful responsibility in having caused her to run the risk. I fear such beauty, among the very poor, is a very dangerous possession."

For many years there was a delightful interchange of visits. I would go down to Oxford for the day to photograph or draw portraits of his child-friends for him. The children would return home, sometimes before, sometimes after lunch; in the afternoon Mr. Dodgson and I would each bury ourselves in a luxurious armchair and talk, or look over sketches I had brought for him to see, or go through part of his vast collection of photographs of famous people taken long years before. Many of Gabriel

Rossetti and his sister Christina, Ruskin,[1] the Millais children,[2] the Terry family.[3]

I remember one of Ellen Terry, taken when she was about eighteen; she happened to laugh at the moment, and she looked one of the loveliest young creatures imaginable.

Then he would make the afternoon tea himself, all in his deft and dainty fashion, and this would bring the happy visit to its close.

A lady friend of Mr. Dodgson's—a good woman, eminently practical and steeped to the lips in conventions—once took me to task.

We had spent the morning at her house sketching her children; at least, I sketched while he talked "Alice."

He left before lunch, and after this meal was over she sent the children away, and, taking up her sewing, sat down in front of me.

"I hear that you spent the other day in Oxford, with Mr. Dodgson?"

"Yes, it was a most delightful day."

"It's a very unconventional thing to do."

"We are both very unconventional."

"Mr. Dodgson is not at all a ladies' man."

"He wouldn't be my friend if he were."

"He is a confirmed bachelor."

"So am I; and, what is more, he is old enough to be my father."

She steadfastly regarded me for a moment, and then said, "I tell you what it is, Mr. Dodgson doesn't think of you as a 'young lady' or anything of that kind, he looks upon you as a sort of 'old child'."

I laughed delightedly. "I don't mind if Mr. Dodgson looks upon me as a sort of old grandmother if only he will ask me down to Oxford."

But I was deeply offended. Our pure and beautiful friendship seemed hurt somehow by this coarse handling.

Sometimes he would come up to town for the day. We would meet, lunch together, and go off to a matinee, sometimes coming back to my rooms for tea. Once I had a ghastly experience.

I have an artist-friend, a rare genius, as unique in his way, perhaps, as Lewis Carroll, but as opposite as the poles. Rugged, full of angles, charm-

---

[1] John Ruskin (1819–1900). Well-known art and literary critic.

[2] Sir John Everett Millais (1829–96). Pre-Raphaelite painter and father of two boys and three girls. Dodgson befriended the family in 1864.

[3] Theatrical family with whom Dodgson was intimate for many years. Including Ellen Terry (1847–1928), Henry Irving's leading lady and one of the great actresses of the Victorian and Edwardian age as well as a close friend of Dodgson; Kate Terry (her sister); and many others.

ing in private life, gauche in the presence of strangers. I call him Solomon Eagle, from his striking resemblance to that picturesque person, and for his holy horror of the world, the flesh, and the devil, particularly as displayed in theatrical performances. I had spoken of this friend to Lewis Carroll, and shown him autotypes of his work which greatly impressed him. One day Lewis Carroll had been up in town and had taken me to a matinee at Wilson Barrett's theatre.[1] We had just returned to my rooms for tea when my artist-friend was announced. I asked permission to introduce him. "Very pleased," was the reply, and Solomon Eagle was ushered in.

Instantly my blood began to curdle, for I saw at a flash what I'd never dreamed of before, that they were denizens of different planets, so to speak.

At the sight of a "grown-up," the "Lewis Carroll" dropped from off my Oxford friend like the petals of a rudely shaken flower, leaving only the dry stalk. He stood, an aristocrat to his finger tips, the cold, serene, dignified Mr. Dodgson, the Don. Solomon Eagle, nervous, gauche, spasmodic.

The greeting passed off well enough. If only it had ended there! Feeling at a loss for the next remark my artist-friend approached the chimney-piece, against which, at either end, Mr. Dodgson and I were leaning. He caught sight of a photograph, which I had propped up against the clock, of the principal actress in the play we had seen that afternoon.

"What on earth possesses you to put a thing like that up there? It will ruin your taste!"

Before I could reply Mr. Dodgson quietly remarked: "Our friend kindly accepted that from me. It is a portrait of the chief actress in Mr. Wilson Barrett's theatre."

The fat was now fully in the fire, and I drew in my breath and braced myself for the inevitable frizzle.

"The theatre is the mouth of Hell!" shrieked Solomon Eagle, and he flashed his brilliant eyes upon me.

There was no help for it, so I meekly replied: "Yes, we've both been there this afternoon."

Like a mad bull he turned to rend another rag, and glared again at the hapless photograph.

---

[1] Wilson Barrett, actor manager. The play was probably *The Golden Ladder* by George R. Sims and Wilson Barrett, which Dodgson and Thomson attended at the Royal Globe Theatre on 11 February 1888. The lead actress referred to below was Miss Eastlake. Dodgson wrote that "It is an excellent play, though too painful to take children to."

"Photography and Ruskin have done more to destroy English art than—."

"Mr. Ruskin is one of my dearest friends, and I am proud of it. Photography is my hobby, and I am not inclined to apologise even for that!" Thus the Don.

I prayed that the earth might open and swallow me up, or that a fiery chariot would descend and whirl away Solomon Eagle on the spot; but Providence didn't see fit to interfere.

Solomon, brought to his senses by this cold douche, instantly apologised, and suddenly remembering a pressing engagement, abruptly wished me "Good-night."

The bows that were exchanged between the artist and the Don were studies in deportment. If I hadn't been so utterly wretched I should have laughed.

I took the ruffled Eagle downstairs to see him safe off the premises, and his language was more picturesque than polite.

"Bloodless fossil!" were the words he hurled into the quiet street as he left the door.

In justice to him I ought to say that he had not the faintest idea beforehand of Mr. Dodgson's views on any subject.

It was entirely my fault in bringing ice and fire together.

On re-entering the room I found that Mr. Dodgson, the don, had disappeared, and "Lewis Carroll" was standing on the hearthrug, smiling.

"I don't like your friend," he remarked airily.

"Neither do I," I groaned, dropping into a chair like a limp rag, "at present."

Oddly enough they collided on my door-step a day or two later, and after a few minutes' chat, Mr. Dodgson came up to my room.

"I met your friend," he said, "and I was mistaken—I think now that I rather do like him."

Mr. Dodgson had a great liking for the theatre, as long as the plays had a wholesome flavour, and he had a special fancy for little girl actresses. He would get to know them and load them with kindness—asking them down to Eastbourne with their mother[1] for a few days or weeks of fresh air. But as soon as they emerged from childhood he gently detached himself. It was a strange characteristic of his, to care greatly for his little friends only so long as they were children. Their chief fascination for him passed from them as they lost their childishness. But to

[1] In fact Dodgson's Eastbourne young visitors were nearly always unchaperoned.

his tender and generous nature all appealed who were in need of sympathy.

I remember once—having offered me his rooms at Eastbourne for a week or two, as he was detained in Oxford—his writing: "If you have any overworked, tired friend who needs a holiday, take her down with you to occupy that other room, and please allow me to pay all her expenses." In small things as well as greater, his kindly courtesy and consideration were unfailing.

"While you are doing so much for my pleasure"—(I was designing some fairies for him)—he wrote once, "it were churlish not to do something for yours, so, though I am writing 'Logic' about six hours a day, I have spared time to write out these two little acrostic poems for you."

"I wish you a very Happy Christmas," began another letter, "but my wishes will come a little late, as, to spare the overworked letter carrier, I shall leave this to be posted after Christmas."

The most distinguishing features about him were his delicate reserve and dignified modesty; he shrank with painful over-sensitiveness from being "lionised" in any way. To be pointed out as the "Alice man" was to him abhorrent.

"I can't imagine what people saw in *Alice*," he once remarked to me. "I believe its popularity has more to do with Tenniel's lovely drawings, than with any nonsense of *mine*!"

"I never read criticisms of my work," he said to me only the other day. "If they praise one it makes one vain: if they abuse one it makes one angry, and both states of mind are bad."

I always had a mysterious feeling, when looking at him and hearing him speak, that he was not exactly an ordinary human being of flesh and blood. Rather did he seem as some delicate, ethereal spirit, enveloped for the moment in a semblance of common humanity.

His head was small, and beautifully formed; the brow rather low, broad, white, and finely modelled. Dreamy grey eyes, a sensitive mouth, slightly compressed when in repose, but softening into the most beautiful smile when he spoke. He had a slight hesitancy sometimes when speaking, that reminded one of dear Charles Lamb's[1] immortal stammer; but though Mr. Dodgson deplored it himself, it added a certain piquancy, especially if he was uttering any whimsicality.

It is just two months to-day since we spent our last day together.

He had written asking me if I would go with him to the matinee of

[1] Charles Lamb (1775–1834). English writer and author (with his sister Mary) of several books for children including *Tales from Shakespeare* (1807).

The Little Minister,[1] and I had suggested that he should come to my studio in the morning for an hour's sketching.

I secured a lovely little girl as model, and he promptly appeared at 11 o'clock.

He was charmed with the child, and they made friends at once. In the "resting" intervals he sat with her on his knee drawing comic pictures to amuse her, and warming her little hands, for the morning was chilly. We were to lunch with some friends of his in Lowndes-square, and being rather behindhand, we hailed a hansom. Something frightened the horse and it bolted with us.

"Well," observed Lewis Carroll serenely, "if we are alive when we reach Lowndes-square, we certainly shall not be late!"

He was exceptionally brilliant that day at lunch, full of repartee and anecdote. He looked extremely well, and as if many years of work still lay before him. As we were driving to the theatre he confessed to me that he had been working very hard lately, sitting up till 5 o'clock in the morning. When I ventured to gently remonstrate, he smiled.

"It suits me," he said; "I feel very well."

Then suddenly turning to me, he said, while a wistful look grew in his eyes, "My time is so short. I have so much to do before I go, and the call might come any day."

Little did I dream that before two months were over the call *would* have come.

He was charmed with *The Little Minister*.

Miss Winifred Emery,[2] by her enchanting personation, won his warmest admiration.

The part of the boy was played by a young actress with whom he had just become acquainted.[3]

While in the theatre he scribbled a note. "May we come round and see you?" and handed it to one of the officials to send in to her.

Presently back came a note; he glanced at the superscription, and on his lips played the old whimsical smile.

"This is evidently intended for *you*," he said quietly. "Will you allow me to open it?"

---

[1] *The Little Minister* by J. M. Barrie (1860–1937), best known as author of *Peter Pan*. Dodgson and Thomson saw the play at The Theatre Royal, Haymarket on 20 November 1897. Dodgson described it as "a beautiful play, beautifully acted. . . . *The Little Minister* is a play I should like to see again and again." It was the last time he attended the theatre.

[2] Actress and wife of Cyril Maude. Maude was one of the managers of the Theatre Royal Haymarket and a lead actor in *The Little Minister*. Emery played Lady Babbie.

[3] Miss Sydney Fairbrother, who played the role of Micah Dow.

It was addressed, "Mrs. Dodgson"!

How we laughed.

"That 'we' was misleading," he remarked. "Well, we are certainly labelled *now*!"

After the play he asked, "Have you anything special to do? If not, come and have tea with me at Paddington, and see me off."

All too swiftly the last half-hour glided away, and I went with him to take his seat.

At the door of the compartment he turned to shake hands. "Good-bye, I've had such a happy day," he said simply. "I hope *you* have."

Just before Christmas he wrote, "Write to me at 'The Chestnuts,' Guildford, where I shall stay until the middle of January."

The middle of January is here.

But the hand that wrote those words has laid down the magic pen for ever; and he, who called forth the purest, gayest laughter the world has ever heard, has passed into that silent land of shadows where no voice can reach, nor footsteps follow!

A grey January day, calm, and without a sound, full of the peace of God which passeth all understanding.

A steep, stony, country road, with hedges close on either side, fast quickening with the breath of the premature spring. Between the withered leaves of the dead summer a pure white daisy here and there shone out like a little star.

A few mourners slowly climbed the hill in silence, while borne before them on a simple hand-bier was the coffin, half hid in flowers.

Under an old yew, round whose gnarled trunk the green ivy twined, in the pure white chalk earth his body was laid to rest, while the slow bell tolled the passing—

"Of the sweetest-soul
That ever looked with human eyes."[1]

Good-Will
March 1898

The news that "Lewis Carroll" was dead brought a touch of real sadness into many a home. His books are so well known—they have taken their place, and that no mean one, in English literature—that

[1] From Tennyson's *In Memoriam*, LVII, 11–12. The same inscription appeared on the wreath which Thomson sent to Dodgson's funeral.

he seemed to many who had not seen him like an old friend. So there came the wide-spread sense of loss when it was known that he had been taken from us. And yet "Lewis Carroll" had been really lost to us for some time; the magic wand was broken, and it was certain that he would never again reach the level of his best work. No doubt in *Sylvie and Bruno* there is much that recalls the old skill and genius, passages here and there that rise to his own peculiar level; but such passages occur in the middle of a great deal that does not reach to the standard he had taught us to expect from him, among much, too, that is almost commonplace. In the *Alices* we are never let down a single moment; the weird, entrancing nonsense carries us along from the first word to the very last. We bubble over with laughter of the purest enjoyment—laughter of which he alone can touch the spring. Others have tried to imitate him, but one and all they have been more or less dismal failures. That wild, impossibly possible combination of sense and nonsense, so delightfully interwoven, that one never knew whether he was in earnest or in fun; whether he was really dealing in his odd way with some of the subjects of everyday life, or whether it was all one delightfully extravagant jest. Different people tried to interpret him, and to find out, for example, what the "Snark" was; but just as they imagined they had seized the meaning of his allusion, they "looked again"[1] (to use his own phrase), and it seemed something wholly different. His allusion had hopelessly eluded them.

The Rev. Charles Lutwidge Dodgson was the son of Archdeacon Dodgson, who was himself a considerable Mathematician and a brilliant scholar.[2] He was educated at Rugby School, and matriculated at Christ Church in 1850. He was nominated about two years later a Student of Christ Church by Dr. Pusey,[3] and in 1854 took his degree, obtaining a first class in the Final Mathematical School, and a third class in the Classical School. He was appointed Mathematical Lecturer of Christ Church in 1855, and held the office till 1881. Since the latter date he had resided continuously at Christ Church; working steadily at many different sub-

---

[1] This is an allusion to Dodgson's poem "The Mad Gardener's Song" from *Sylvie and Bruno*. Each stanza takes the form of three couplets, the first beginning "He thought he saw . . .," and the second beginning "He looked again . . ."

[2] Though Dodgson's father did study mathematics, obtaining First Class Honours, this statement is slightly misleading. Most of Archdeacon Dodgson's publications were of a religious, not a mathematical, nature; they included a translation of Tertullian for *The Library of Christian Fathers* and numerous sermons.

[3] Edward Bouverie Pusey (1800–82), Canon of Christ Church and Regius Professor of Hebrew was a fellow undergraduate at Christ Church with Dodgson's father, and the two were close friends.

jects, principally interested in what seemed to many the dry bones of formal logic, and the logical bases of Euclidean geometry. In the latter subject, it may be remarked in passing, he did some excellent work.

The character of "Lewis Carroll," like his books, is difficult to define or explain. It seemed made up of such very contrary elements. Two characteristics, however, stand out very prominently; they also may appear, at first sight, not to be very closely connected; really in him they were very near together, and each explained and supplemented the other:—his deep and true reverence, and his great affection for children—or, rather, it ought to be said—for little girls; boys never attracted him,[1] in spite of his charming little boy-creation, Bruno.

Reverence for his deeply reverent nature forbids that much should be said about him in this regard; only this aspect of his character must never be left out of sight, otherwise the picture formed of him will be very incomplete. He sternly silenced any attempt, to use with the slightest touch of jest, sacred words,—words that is from the Bible, or usually associated with sacred things; no doubt he sometimes seemed pedantic or puritanically exacting, but those who knew him best, thank him for the stand he made against irreverence, flippancy or idle jesting in regard to sacred things, in an age which, whatever its peculiar graces may be, certainly fall very far short in this respect.

His love of children was remarkable; with them he became as a child. Of course he had puzzles and games and so forth to show them, but his relations with children did not stop at this point, as they usually do in grown-up people. He so entered into and understood their ways, that children would talk as freely and readily to him as they did to each other; and in this way he came actually in contact with their minds,—with their odd, sometimes unexpectedly serious, ways of looking at things. It is this that makes his books so real. Many of the remarks found in them are things actually said by children to himself. His ear was always open and ready to catch any hint that he could get in this way; and such chance words and phrases were always carefully noted, and sooner or later worked up for one of his books. For these wonderful books were not written straight on end. A regular medley of odd bits were got together,[2] poems,

---

[1] Although Dodgson himself once said, "I like children, except boys," he certainly was friendly with some boys, including young Harry Liddell, Alice's brother. The vast majority of his child friends, though, were girls.

[2] In the Preface to *Sylvie and Bruno*, Dodgson wrote "As the years went on, I jotted down, at odd moments, all sorts of odd ideas and fragments of dialogue that occurred to me. . . . And thus it came to pass that I found myself in possession of a huge unwieldy

stray remarks, sometimes whole scenes; and then, by some slow process, almost unintelligible to others, the complete book would gradually emerge, and he would defy critics to find out when and how he had pieced it together. The *Snark* certainly was built up from a single line that came into his head one day as he was on a walk, and which he at once noted down in his pocket-book. It is the last line of the poem "The Snark was a Boojum, you see."

No one can doubt that this insight into childlife, the power of "becoming as a little child,"[1] was the outcome of the singular purity and the deep and reverent religious feeling that formed the basis of his character. Of late years he had more and more withdrawn himself from society, and lived in his own way, very much apart and by himself; but those who knew him well, know that their life is far poorer for his loss; that they must always miss a friend whose place no other can ever take—who united to the gifts that are usually associated with friendship, the attraction of a quite unique personality enriched by true genius—genius which, while they are sure of its reality, they frankly say they find it hard to define. He has passed from their sight, but they are sure that in the mercy of God he has received the blessing that waits for pure, simple, and child-like lives: "Blessed are the pure in heart, for they shall see God."[2]

The Graphic (London)[3] January 22, 1898

The death of the Rev. Charles Lutwidge Dodgson, better known as "Lewis Carroll," the author of *Alice's Adventures*, will be received with feelings of regret wherever English is spoken. Mr. Dodgson was born in 1833, and was educated at Christ Church, Oxford, where he took a first-class in the Mathematical Schools in 1854, became senior student of Christ Church soon afterwards, and took orders in 1861. From 1855 to 1881 Mr. Dodgson was mathematical lecturer at his college, and published several books on higher mathematics. In 1865 his *Alice's Adven-*

---

mass . . . which only needed stringing together." This was not his only method of writing, however, as much of the first *Alice* book was told extemporaneously.

[1] Probably a reference to 2 Kings 5:14, "and his flesh came again like unto the flesh of a little child."

[2] Matthew 5:8.

[3] This obituary was accompanied by the photograph of Dodgson by W. Shawcross of Guildford.

tures in Wonderland, written to please Dean Liddell's children, took the world by storm. The Queen, it is said, was so pleased with the book that she ordered that the next book from the same author should be sent to her, and accordingly, *An Elementary Treatise on Determinants* was forwarded by the publishers to Her Majesty. *Alice's Adventures* was followed by *Through the Looking Glass* in 1871, which was, however, not so successful, though exceedingly clever. The *Hunting of the Snark*, *Alice's Adventures Underground*, and *Sylvie and Bruno* came also from Mr. Dodgson's versatile pen, but the reputation of "Lewis Carroll" was already too high to be increased by these works, witty and humorous as they were.

The Guardian[1] January 19, 1898

Though no longer children in years there were many to whom "Lewis Carroll" still gave this title: "I make very few new child-acquaintances now, but I have still so many old child-friends, some of them married women with children of their own," was a remark he made to me recently, with one of those delightful smiles which usually accompanied such sayings. It is, therefore, in the capacity of one of his oldest "child-friends," and one who as such saw perhaps more of him in the last few years than most others, that I venture to give some recollections of our "Mr. Dodgson," for the sake of those who knew him less intimately.

"Lewis Carroll" is known almost throughout the world by his incomparable *Alice*, and yet in many respects that author and the Don of Christ Church, Oxford, were two distinct persons, and he preferred that the public should act as if this were the case. "I claim no authorship with any such book," was his invariable answer to letters that came addressed to "the Rev. C. L. Dodgson" on the subject of his fictitious works; while, on the other hand, the unfortunate people who sent any communications to "Lewis Carroll, Christ Church, Oxford," had their notes returned unopened on the plea that no such person lived at that address. It was his horror of being lionised which made him thus repel would-be acquaintances, interviewers, and the persistent petitioners for his autograph, and

[1] This article was signed "By a child-friend, B.H." and was written by Beatrice Sheward Hatch (1866–1947). Dodgson was close friends with Hatch and her family for over two decades. Edwin Hatch, father to Beatrice, was Vice-Principal of St. Mary Hall, Oxford.

he would afterwards relate the stories of his success in silencing all such people with much satisfaction and amusement. Not long ago Mr. Dodgson happened to get into correspondence with a man whom he had never seen, on some question of religious difficulty, and he invited him to come to his rooms and have a talk on the subject. When, therefore, a Mr. X. was announced to him one morning he advanced to meet him with outstretched hand and smiles of welcome. "Come in, Mr. X., I have been expecting you." The delighted visitor thought this a promising beginning and immediately pulled out a note-book and pencil, and proceeded to ask "the usual questions." Great was Mr. Dodgson's disgust! Instead of his expected friend here was another man of the same name, and one of the much-dreaded interviewers, actually sitting in his chair! The mistake was soon explained, and the representative of the press was bowed out as quickly as he had come in. On another occasion the editor of a collection of biographies wrote to ask for the honour of including his in the forthcoming volume. Mr. Dodgson would not even allow him the satisfaction of an autograph letter in reply, but sent over to the present writer of this paper to request that a copy of his refusal should be made and forwarded by her.

But to his "child-friends" he was the most gentle, affectionate, and sympathetic of companions. It is hardly necessary, perhaps, to say they were always of the weaker sex. He did not like boys any more than he liked *very* young children, though it is difficult to believe that, after the delightful creation of "Bruno." But girls of all sorts interested him—old and young, plain and pretty; from princesses, whose photographs and letters to himself he would show with some pride, down to lowly born children in a poor parish, where a sweet face would attract him from the band of hope or school audience assembled to hear him tell stories and draw pictures on a blackboard for their special amusement. Many were the friends he made, too, among stage-children, to whom he showed the greatest kindness. The theatre had a great fascination for him, though never would he countenance any play that was not strictly moral and refined. And if a child happened to be taking part in the performance, one of his books, accompanied by a note or card, was pretty sure to find its way round to the stage-door next morning addressed to the little actress.

It is difficult to realise, even while the bell is tolling in our ears this sad morning,[1] that Oxford will no longer see that thin, upright figure

[1] According to S. A. Warner, *Oxford Cathedral*, "on January 18, 1898, a peal of Kent Treble Bob Maximus, 5,040 changes, was rung in three hours, fifty-eight minutes, muffled

walking rapidly through the streets, seeing no one, greeting no one, intent on his own thoughts, alone, or with a child-friend as companion, as he fetches her to have tea in his rooms, or takes her home again. He very rarely associated with any men of his own standing, beyond dining in hall, and of late years he refused all invitations into society. A rule once made must never be broken was evidently a motto of his. I have known him reply to an invitation for a particular day, "Because you have *asked* me, therefore I cannot come, but I will come the next day;" and those who knew of this rule were careful not to *invite*, but merely to give information that they would be at home on a certain day at a certain hour, and leave it to him whether he would come or not. Unless it were understood that no other guests would be there he would even then not come. But what of his own "parties"?—for he was fond of entertaining, and a truly excellent host. The number of guests invited seldom exceeded *one*, though I have known two and even three. He preferred to have his friends singly, and in the case of two sisters whose mother wished them to come together he has been known to decline the pleasure of their company altogether. At one time it was always "a walk and tea" to which one was invited, and this grew into "dinner parties." Never lunch, for he did not take it, and he liked to work uninterruptedly all the morning until nearly two o'clock. But now the last is over—the climb up the winding, wooden staircase in the corner of Tom Quad; the unlocking of the great oak door with "the Rev. C. L. Dodgson" painted in white letters over it; yet another small door to be opened, and there we are in the large familiar room, with its huge windows overlooking St. Aldate's.[1] Under the bookcases are the cupboards which I loved as a child, for besides the mystery of their being unlocked by a special key on Mr. Dodgson's own bunch, they contained such delights in the way of musical boxes, a brown bear that walked when it was wound up, and packets innumerable of photographs, held together by india-rubber bands, and catalogued with exquisite neatness. It is many, many years since Mr. Dodgson gave up this amusement, but I shall never forget the delight of mornings spent with him dressed up as a Japanese or a gipsy, or a beggar-child, for my photograph to be taken; watching the mysterious charms that he poured upon them from little bottles that emitted such a curious odour, and in the intervals the running out on the flat roof of the college.

---

in memory of Rev. C. L. Dodgson." Warner also states that Dean Liddell died while this peal was being rung.

[1] Dodgson occupied this large suite of rooms at the northwest corner of Tom Quad from 1868 until his death.

But let us go on to more recent years, since "dinner parties" began. You are seated in a corner of the red sofa in front of the large fireplace, over which hang some painted portraits of his child-friends. Your host tells you that the tiles round the fireplace represent the story of *The Hunting of the Snark*, and he forthwith explains them to you.[1] One anecdote succeeds another, or you are shown the piles of MSS. on which he is at work. Logic is still, as it always has been, his great subject, and Symbolic Logic, of which Part I. was published in February, 1896, the particular branch. Indeed, he would almost make you believe that this study was an essential of life, and was glad of any opportunity to impart instruction to his girl-acquaintances. When dinner is announced you are led into another room, much smaller than the first, and you may be quite sure you will never get the same menu that you had last time; for, besides other registers, lists and catalogues of letters, books, &c., Mr. Dodgson keeps a list of the dishes supplied every time he has guests at his table, and is careful always to look this up when he invites you, that you may not have the same thing twice. Two courses, and only two, is another of his rules—meat and pudding—and no one could wish for more. After dinner follows music in the other room. Mr. Dodgson has no piano, neither does he sing, but he has an organette which plays tunes off perforated cards. To soften the sound this lives in a wooden baize-lined box, with a hole in the side for the handle. It was delightful to watch the expression of that keen face and sensitive mouth as the author of *Alice* solemnly turned the handle, and played tune after tune, regulating the sound by opening or closing the lid, according as he wished to make it louder or softer. Then there are games of his own invention, "Mish-mash," "Landrick,"[2] and so on; also his own Memoria Technica[3] to be explained and possibly learnt. And so the evening slips away, and it is time to be escorted home again.

I have never heard Mr. Dodgson preach, and shall always regret it. It was but rarely that he did so; not that I believe his stammer was any

---

[1] In the 1880s Dodgson had fireplace tiles designed by William De Morgan (1839–1917) installed around the fireplace in his sitting room. De Morgan was an artist and later a novelist whose tiles were popular examples of the Arts and Crafts movement. Dodgson chose tiles which pictured fantastical creatures and a three-paneled sailing ship.

[2] *Mischmasch* (1882) and *Lanrick* (1879) were word games invented by Dodgson and published by him.

[3] A system of mnemonics invented by Dodgson (adapted from a system invented by Dr. Richard Gray in 1730). Dodgson published two cyclostyled versions of *Memoria Technica* (1877, 1888).

hindrance to him in this. He has often told me that he never wrote out his sermons. He knew exactly what he wished to say, and completely forgot his audience in his anxiety to explain his point clearly. He thought of the subject only, and the words came of themselves. Looking straight in front of him he saw, as it were, his argument mapped out in the form of a diagram, and he set to work to prove it point by point, under its separate heads, and then summed up the whole. He took a serious view of life, and had a very grave vein running through his mind. The simplicity of his faith was very beautiful, though he had his prejudices, and some rather curious ideas on religious subjects. But he was liberal to all men, and always ready to help where he could. I have known him take a great deal of trouble over a very small matter, in order to aid or give pleasure to a friend, or even a friend's friend. We shall miss him for a long time to come—not for his books, for his storytelling days were past, but for himself. The well-known bright smile whenever we met; the long calls, when one felt oneself a child again for old sake's sake and life one vast holiday; the familiar and characteristic handwriting in the frequent and amusing notes; above all, the true affection that grows scarcer in these latter days—these are now things of the past, and we mourn.

But the lover of children himself "as a little child"[1] is reaping his reward of rest and peace beyond the veil.

The Guardian
January 19, 1898

Professor Sanday, preaching at Christ Church on Sunday morning, prefaced his sermon with the following reference to the deceased:—

"It falls to me to speak of the unexpected and saddening news which reached us only yesterday, and which casts its shadow over the approaching term. The term will come, but we shall no more have among us the living presence of one who, for all the retirement in which he lived, had for many years made Christ Church his home, and could not help filling no small place in the picture which the world without formed to itself of Christ Church.

"It falls to me to say a few words, and I can only do so as a late comer,

[1] Mark 10:15. "Whosoever shall not receive the kingdom of God as a little child, he shall not enter therein."

who cannot claim the privilege of close intimacy. I do not doubt that my deficiency in this respect will be abundantly made good later, and so the burden of responsibility is for me happily lightened.

"But after all, was any lengthened intercourse necessary to feel that one knew Charles Lutwidge Dodgson? The outer man was familiar to most of those who frequent our streets and ramble through the lanes of the country round. The inner man, in part, lies patent in many a charming book, and in part was so transparent that even the ordinary meetings of daily life, with an occasional walk and talk, was enough to reveal its main outlines. Alas! only these—because all that makes the individual, the infinite play of fancy and the subtle undercurrents of serious and chastened thought, must needs be lost to us; they go with him whose they were to inhabit another sphere than ours.

"The world will think of 'Lewis Carroll' as one who opened out a new vein in literature, a new and a delightful vein, which added at once mirth and refinement to life. Was not this much to have done? Does it not bespeak an original and creative gift of a high order, and of great value in the eyes of those who have the best interests of the nation at heart? 'Whatsoever things are lovely, whatsoever things are pure, whatsoever things are of good report; if there be any virtue, and if there be any praise'[1]—do not these words describe what we associate with this name? Books wholly without spot or stain in the midst of so much that is the reverse of stainless—and all seasoned with such a wholesome salt of humour as won for them a wide popularity. May we not say that from our courts at Christ Church there has flowed into the literature of our time a rill, bright and sparkling, health-giving and purifying wherever its waters extend? Perhaps there are few happier spots in the English literature of the century than its books for children. And how large a share in this is due to him who is gone?

"For this all England is grateful, nay more, all readers of the English tongue. But we in this place have much more of which we are thinking. We know how fully the man bore out the promise of his books. We know what an interest he took in an art which offers great scope for genius and great capacity for good, though beset with not a few temptations. We know what a high and scrupulous standard he brought to this, and how excellent was his influence in regard to it. We know his fondness for children and what trouble he took to make them happy and better. But most of all we know what was the fount and spring from which all these

[1] Philippians 4:8.

varied activities took their direction. We know how behind them all there lay a deep background of religion—a religion severely quiet and retiring, like his character; a religion almost of the closet, after the pattern of the Gospel, 'Thou when thou prayest enter into thy closet, and when thou hast shut the door pray to thy Father Which is in secret.'[1] We know that his public utterances were rare but striking in their rarity. We know how he chose out by preference the smallest and least frequented of our services, and we know how when there the bowed head and close concentrated attention spoke of a soul really communing with its God.

"From those things we may not further lift the veil. He is now in the presence of that God Who seeth in secret, and we may be sure that he has his reward—the reward of a life, perhaps, of even more far-reaching beneficence than we suppose. Do I say "reward'? It is a word of which I feel sure that, in his humility and Christian sincerity, he would have disapproved. I mean that acceptance for which we hope through Jesus Christ our Lord."

Hawaiian Gazette (Honolulu)
January 25, 1898

London—The Rev. C. L. Dodgson, whose nom de plume is Lewis Carroll, author of *Alice in Wonderland*, is dead.

The Illustrated London News[2]
January 22, 1898

The exquisite humorist who was known to the world as "Lewis Carroll," and to a much smaller circle as the Rev. Charles Lutwidge Dodgson, has died at the age of sixty-five. Mr. Dodgson, whose father was Archdeacon of Richmond, was educated at Rugby and Christ Church. He became a grave Oxford Don, and a profound mathematician; published in 1860 a *Syllabus of Plane Algebraical Geometry*, which

---

[1] Matthew 6:6.

[2] On 4 April 1891, *The Illustrated London News* published one of the few biographical pieces on Dodgson that appeared in his lifetime. The short article described his rooms in Christ Church and identified Dodgson as Lewis Carroll, but included some inaccuracies. The present article is accompanied by a detail of the photograph of Dodgson by W. Shawcross of Guildford (printed at the wrong angle).

was followed in 1864 by *A Guide to the Mathematical Student.* From these sober studies Mr. Dodgson burst suddenly into a humorous vein, so quaint, fantastic, and original that *Alice's Adventures in Wonderland* at once attained a popularity it has never lost. It was followed in 1872 by *Through the Looking-Glass*, a further instalment [*sic*] of Alice's experiences. Then came an interval of mathematics, which apparently furnished the inspiration for *The Hunting of the Snark*, and other sprightly sallies which, if not equal to the first venture, have much of the author's genuine quality. The last was *Sylvia and Bruno*, which contained a remarkable preface, written by the Rev. Charles Lutwidge Dodgson, M.A., not by "Lewis Carroll." It appeared to be a kind of penitential homily on the frivolity of fun and on the impropriety of allowing children to go to the theatre.[1] In his time Mr. Dodgson had been an enthusiastic play-goer, and the stage version of *Alice's Adventures* has delighted both old and young. But in the closing years of his life Mr. Dodgson was once more the Oxford Don, resentful of messages addressed to Lewis Carroll, but, as his friends have testified, full of kindness and humour under a prim, mathematical exterior. Few English writers have contributed so much to the gaiety of a whole people.

The Independent (London) January 20, 1898.

Though "Lewis Carroll's" life work was devoted to the higher mathematics his name will always be remembered as the author of the most popular child's book of the century, and as the man who opened out a new and delightful view in literature. *Alice in Wonderland* and *Through the Looking-glass* were written for children, but they have been a source of infinite amusement for grown-ups; and on the very day of the author's death a passage from *Alice* was quoted in all the newspapers as a commentary upon the military scandal which is agitating Paris. *Alice in Wonderland, Uncle Remus*[2], and *The Golden Age* are a trio of books for children which have the stamp of immortal popularity. "Lewis Carroll" was

[1] On the contrary, Dodgson continued to enjoy taking young children to the theatre until the end of his life. He does argue, in this preface and elsewhere, against attending any play of immoral content or irreverent nature.

[2] *Uncle Remus, his songs and his sayings* (1880), by Joel Chandler Harris (1848–1908).

a clergyman and an Oxford Don, but above all he was a writer of children's books. He was grave and gay by turns, and could mingle grave thoughts with frolicsome fun. At one time he used to issue a fly-leaf to "every child who loves Alice." This was one of them:—

> God does not mean us to divide life into two halves—to wear a grave face on Sunday, and to think it out of place to even so much as mention Him on a week-day. Do you think He cares to see only kneeling figures and to hear only tones of prayer, and that He does not also love to see the lambs leaping in the sunlight and to hear the merry voices of the children as they roll among the hay? Surely their innocent laughter is as sweet in His ears as the grandest anthem that ever rolled up from the 'dim religious light' of some solemn cathedral! And if I have written anything to add to those stories of innocent and healthy amusement that are laid up in books for the children I love so well, it is surely something I may hope to look back upon without shame and sorrow (as how much of life must then be recalled) when my turn comes to walk through the valley of shadows![1]

In a different vein was the letter he addressed to a child of eight, to whom he was sending a book. Then he wrote:

> My Dear Magdalen,—I want to explain to you why I did not call yesterday. I was sorry to miss you, but you see I had so many conversations on the way. I tried to explain to the people in the street that I was going to see you, but they wouldn't listen; they said they were in a hurry, which was rude. At last I met a wheelbarrow that I thought would attend to me, but I couldn't make out what was in it. I saw some features at first, then I looked through a telescope, and found it was a countenance; then I looked through a microscope, and found it was a face! I thought it was rather like me, so I fetched a large looking-glass to make sure, and then, to my great joy, I found it was Me. We shook hands, and were just beginning to talk, when Myself came up and joined us, and we had quite a pleasant conversation. I said: 'Do you remember when we all met at Sandown?' And Myself said: 'It was very jolly there; there was a child called Magdalen.'

---

[1] From *An Easter Greeting to Every Child Who Loves "Alice"* (1876). See above p. 6n.

> And Me said: 'I used to like her a little. Not much, you know—only a little.' Then it was time for us to go to the train—and who do you think came to the station to see us off? You would never guess, so I must tell you. They were two very dear friends of mine, who happen to be here just now, and beg to be allowed to sign this letter as your affectionate friends, Lewis Carrol [*sic*] and C. L. Dodgson.

The death of Mr. Dodgson, of Christ Church, has cast a shadow over Oxford, but the removal of "Lewis Carroll" has struck sorrow into many a young heart all over the world.

The Independent (New York) January 20, 1898

The death of "Lewis Carroll," the Rev. C. L. Dodgson, which occurred last week, will seem almost like a personal loss to those of middle life who have not yet forgotten the charm that *Alice's Adventures in Wonderland* and *Through the Looking-Glass* had for them in the late sixties and early seventies. *Literature*, in its Christmas issue, pays the following tribute to "Lewis Carroll," and justly accredits him with having "created" a new genre in literature[1]

> in which humor and even satire adapted to the taste of the mature intelligence is cunningly combined with those elements of the romantic and marvelous which form the delight of childhood. The humorist and scholar who has won a lasting reputation under the pseudonym of 'Lewis Carroll' was undoubtedly the inventor, as he has been the most successful producer of this species of literature; and the author of *Alice in Wonderland* and *Through the Looking-Glass* has done more than produce that host of more or less successful imitators by whom all inventors are followed. He has, no doubt, done much during his well-nigh thirty years of popularity to create and develop that taste among his youthful public which has ever since found so many to cater for it. It has, at any rate, been mainly his work to revolutionize the 'wonder-story' for children who have outgrown actual in-

[1] The quote is from "Christmas Books," *Literature*, Vol. 1 no. 10 (25 Dec. 1897), 289–90. Unsigned.

> fancy; and to have set the fashion, now so firmly established that the more intelligent of young readers would almost resent any departure from it, of providing them with literature which they observe, no doubt with pride, to be as popular with 'grown-ups' as with themselves. This discovery of a fresh field of common interest, and a new bond of comradeship between the child and the adult, is on the whole, of course, to be welcomed.

It is as a writer of children's books rather than as a mathematician or a divine that his name will be remembered.

The Isis (Oxford)
January 29, 1898

While Christ Church mourns the loss of her venerable Dean, she grieves also for the death of that tender spirit whose *Alice in Wonderland* was inspired by the Dean's daughter. The author of the most delightful of children's books was a clergyman and a mathematical don of great learning. The gentle features and kind heart of the Rev. C. L. Dodgson, whose name will go down to posterity as "Lewis Carroll," were known of few, but of all who knew him he was loved. His shrinking, tender spirit shunned the stress and uncouthness of the world; in his ways he was representative of an older generation of Oxford scholars. By those who knew him he will be much mourned.

Isle of Wight Chronicle[1]
January 20, 1898

The Rev. Charles Lutwidge Dodgson ("Lewis Carroll") died on Friday, at the residence of his sisters, The Chestnuts, Guildford. He was born in 1833, and, entering Christ Church, Oxford, graduated B.A. in 1854, with a first-class in mathematics. He was elected a student of his college, took orders in 1861, and from 1855 until 1881 was mathematical lecturer. In 1865 he published *Alice's Adventures in Wonderland*, a humourous fairy story on the common surroundings of child-life, which, with its continuation, *Through the Looking-glass* (1871), has been

[1] A virtually identical piece appeared in the Rugby *Gazette and Times* (20 Jan. 1898).

translated into most of the languages of Europe. The Rev. gentleman also published *Phantasmagoria* (1869), *Hunting of the Snark* (1876), *Euclid and his Modern Rivals* (1879), *Rhyme? and Reason?* (1883), *A Tangled Tale* (1886), *Game of Logic* (1887), *Sylvie and Bruno* (1889), *Curiosa Mathematica* (1888), and *Symbolic Life* (1896).

The Jamaica Post (Kingston, Jamaica) January 18, 1898

The death is announced of the Revd Dodgson, the author of "Alice in Wonderland."

Lady's Pictorial (London)[1] January 27, 1898

Nurseryland is in mourning, and even children of a larger growth are lamenting the loss of one who has been a valued friend of their youth. Lewis Carroll is dead, and it is not too much to say that there has passed from among us a writer whose place it will be impossible to fill. To a couple of generations of little folks, the author of those delightful books, *Alice in Wonderland* and *Through the Looking Glass*, has been indeed

A Fairy Godfather.

He has probably given more pleasure to children in every part of the globe than any other person of the century. There is surely no little maiden in the wide world who has made the acquaintance of dear little Alice, the White Rabbit, the Duchess, and the Dormouse, who has not felt a deep and lasting affection for them. Some of us, nay, thousands of us, who are now fathers and mothers, remembering the hours of joy we spent in company with our treasured Lewis Carroll volumes, feel the keenest satisfaction in reading them over again to our own little lads and lasses. For us the fun of "Father William" is just as fresh as it is to our little listeners. Time has not withered nor custom staled the infinite variety of this fascinating writer's humour; and, indeed, it was but the day before his death that the *Times*, apropos of the Dreyfus trial, quoted a long passage from

[1] This article, "A Wizard of Wonderland," was signed "Vera."

the trial of the peccant "Knave of Hearts," which grown-ups read with the liveliest amusement.

To have written two books for children which have become English classics in itself entitles the Rev. Charles L. Dodgson—or, as we shall all still love to call him, "Lewis Carroll"—to an honoured place among the writers of the century. So long as there are children to read, his name will be a household word, and he, who was in life so true a lover of children, being dead, will still speak to them for all generations through those delightfully nonsensical rhymes and fascinating whimsical creations, in which both the grown-ups and the little people find unending delight. It is thirty-three years since the gates of Fairyland were thrown open again by Mr. Dodgson, and his truly wonderful Wonderland was first explored by delighted children. The small boys and girls who then listened with wide-open eyes and breathless excitement to Alice's adventures, and learned the now familiar "Walrus and the Carpenter" and "Beautiful Soup," have since revelled in the *Hunting of the Snark* and *Phantasmagoria*; but it is always with a feeling that they are renewing their youth and turning awhile to the society of a dear old friend that they pick up the *Alice in Wonderland* and *Through the Looking Glass* that now find their place on the bookshelves of their own children's nurseries, and it is safe to say that they will be just as fresh to them when their grandchildren come to an age to appreciate the treasure-house of fancy and fun built up for them by Lewis Carroll.

Since these precious possessions of our childhood's days first appeared there have been imitators without number of their author's style; but, clever as many of the books have been which are obviously modelled upon the works of Mr. Dodgson, the writer has not yet appeared who has outrivalled or even equalled him as a story-teller to children. To him, undoubtedly, belongs the honour of having

Turned the Tide of Nursery Literature

into its present channel. Before Alice went to Wonderland and journeyed through the Looking Glass the books that were written for children were positively appalling in their dulness [*sic*]; but when Alice came all was changed. It was not possible, of course, for all to people the realms of fancy with such delightful and immortal creations as the March Hare, the Hatter, the Mock Turtle, and Tweedledum and Tweedledee. It was not possible for them all to place themselves, as Lewis Carroll did, in direct sympathy with every little reader in every rank of life, for the ragged urchin who obtains these volumes from an East-end public library enjoys them as heartily as the grandchildren of our Queen; but they learned from him how to write for children, and it is undoubtedly to his example and influence that the charming story books that now flood the book market owe their origin. There may be a tendency, perhaps, to over-elaborate the books now produced for the little ones. They are too richly bound, maybe, and sometimes they are inclined to be a little over-subtle in their humour; still they are an immense improvement in tone and style upon *Melbourne House*, *Queechy*,[1] *The Parent's Assistant*,[2] *Sandford and Merton*,[3] *Frank*,[4] and their immediate successors. No one will ever write again for the nursery as they did before Lewis Carroll gathered the children around him, and poured into their delighted ears the fanciful, humorous, and simple fairy tales that have made him the very Prince of nursery story-tellers. In private life he was the most charming companion, and his witty sayings are almost household words in Oxford, where he was for many years a professor of mathematics. Over the children who were lucky enough to enjoy the privilege of his personal acquaintance he exercised a positive spell. They would have followed him anywhere, as though he had been a second Pied Piper; the naughtiest boys and girls became good when they were with him; the model little people became positive angels without wings, and there was no treat and no pleasure that they would not gladly have forfeited to spend an afternoon in the rooms at Oxford where so many hundreds of children have passed hours of sweet delight in his company. He himself would have wished no better epitaph than

"He Loved and Was Beloved By All Children,"

and surely to have brightened, as he has, the hours of childhood, and to have enshrined himself in the hearts of little children, is to have lived and died happily. A memorial for all time will exist of him in our nurseries, but it would be a charming idea if every child who has loved his creations would honour his memory in a practical form by contributing a small sum toward founding a ward or a cot in the Great Ormond-street Hospital for Children, to be called after the name by which he endeared himself to his little readers, or, perhaps, even the "Wonderland" Cot.[5]

---

[1] *Melbourne House* (1864) and *Queechy* (1852) were novels by Susan Bogert Warner (American 1819–85) which she published under her pseudonym Elizabeth Wetherell.

[2] 1796 collection of children's stories by Maria Edgeworth (English, 1767–1849).

[3] Collection of stories by Thomas Day (English 1748–1789), published in three volumes in 1783, 1786, and 1789.

[4] Novel for children by Jacob Abbott (1803–79), American author of many books for children, including the Rollo series.

[5] A Lewis Carroll Cot was in fact established at the Great Ormond Street Hospital for Children (see above pp. xiv–xviii).

Longman's Magazine (London)[1]
March 1898

That the death of the author of *Alice in Wonderland* must be greatly regretted is an example of the obvious, worthy of my ingenious contemporary. Nobody since Dickens has given us so many popular sayings and left his mark so firmly in common speech as Mr. Lewis Carroll. How well one remembers, after more than thirty years, the happy surprise which Alice brought! The entire novelty of the amiable non-sense! No one could imitate it, though many still try the impossible feat. Probably the same author's remarks on the new belfry of his college are now very 'scarce,' like the books in Mr. Pearson's new catalogue.[2]

Llandudno Advertiser (Llandudno, Wales)
January 20, 1898

Lewis Carroll, whose real name was Charles Lutwidge Dodgson, and who was the author of *Alice in Wonderland*, died on Friday last at Guildford. The celebrated and charming book, which attained a world-wide popularity, was written to amuse Miss Alice Liddell, of Penmorfa, Llandudno,[3] during an illness from which she was suffering in the summer of 1865.[4] Old residents here well remember when Llandudno was one of the favourite resorts of the aristocracy, when the Dean of Christ Church, Oxford, with Mrs. Liddell and daughters, were the cynosure of all eyes at Church Parade on Sundays, which was then a miniature of what Hyde Park is at the present day.

---

[1] This article was signed "Andrew Lang," (1844–1912), British journalist, poet, and folklorist. Author of a series of books of fairy tales.

[2] The remainder of this article is a discussion of the items in Mr. Pearson's rare book catalogue.

[3] The Liddell family spent nine summers at Penmorfa, their house in Llandudno, on the north coast of Wales from 1863–71. There is still controversy over the question of whether or not Dodgson ever visited them there. Local tradition holds that "The Walrus and the Carpenter" was composed there, but there is no concrete evidence to support this claim.

[4] The Alice story was first written in 1862 (see above p. 6n) and was published in the summer of 1865.

Lloyd's Weekly Newspaper (London) January 16, 1898

The Rev. Charles L. Dodgson, M.A., whose writings have gained him world-wide fame under the pen name of Lewis Carroll, died on Friday afternoon, at The Chestnuts, Guildford, the residence of his sister. Mr Dodgson was senior student and mathematical lecturer at Christ Church, Oxford, and when the famous fairy-book *Alice in Wonderland* appeared in 1865, very few suspected that the grave and reserved mathematician was its author. It was this book that the Queen was so delighted with on reading it for the first time, some ten years after its publication. In a graceful communication her Majesty thanked the author for the pleasure he had afforded her, and requested that his other book, *Through the Looking Glass*, published in 1871, might be sent to her.[1] Mr. Dodgson's other well-known, whimsical books are *Phantasmagoria and Other Poems*, 1869; *Songs from Alice in Wonderland*, 1870; *The Hunting of the Snark*, 1876; *Doublets: A Word Puzzle*, 1879; *Rhymes and Reason*, 1883; *A Tangled Tale*, 1885; and *Alice's Adventures Underground*, 1886. He also wrote many solid and serious works on Mathematics and Logic, contributing to the latter science a curiously original work for learners *The Game of Logic*, which was published in 1887. The deceased author was noted for his very great love for children. He was in his 66th year.

The London News January 20, 1898

The funeral of that whimsical genius whom all the world knows as "Lewis Carroll" was wholly in harmony with the simplicity which distinguished him when alive. "When my turn comes to walk through the valley of shadows, let there be no palls, no plumes, no pomp." And no hearse, no horses. Some such directions as these were found in his rooms at Oxford, and his sacred wishes were faithfully executed by his brothers and sisters. It would be impossible to conceive a more exquisitely simple service, or a more impressive one. And yet the last offices

---

[1] No evidence exists to support this claim, and it is probably a corruption of the more famous, and equally untrue, Queen Victoria anecdote (see above p. 18n).

were performed in no solemn cathedral, to the air of no rolling anthem, with no dim halo of religious light.

At about half-past eleven o'clock yesterday morning the procession from the old red brick house in which Mr. Dodgson died emerged from the ancient Castle Arch into one of the many quaint byways with which Guildford abounds. A plain oaken coffin, covered with beautiful flowers, and drawn on a hand carriage, was the centre of a small group of mourners, who soon reached the gate of the old church of St. Mary's,[1] not much more than a hundred yards down the road, to the tolling of the bell in the square old tower. Winding its way down the grave-lined path of cobbles, the body was met by the Dean of Christchurch, Dr. Paget,[2] and the Rector, Mr. Grant, and thence carried into the nave of the church, in which a handful of people were already seated, to the solemn words of the Office for the Dead: "We brought nothing into this world, and it is certain we can carry nothing out. The Lord gave, and the Lord hath taken away; blessed be the Name of the Lord." Then rose choir and congregation, and sang:

> The Saints of God! their conflict past,
> And Life's long battle won at last.
> No more they need the shield or sword,
> They cast them down before the Lord.
> O! happy Saints! for ever blest,
> At Jesus' feet now safe you rest.
> The Saints of God! their wanderings done
> No more their weary course they run,
> No more they faint, no more they fall,
> No foes oppress, no fears appal;
> O! happy Saints, for ever blest,
> In that dear home how sweet you rest.

The beautiful service, undertaken partly by the Dean, partly by the Rector, was soon ended; there was another hymn; then once more was the body carried out of the venerable interior into the soft and balmy air

---

[1] This was the church which Dodgson attended while at Guildford. He also preached there on occasion (see above p. 5n).

[2] Francis Paget (1851–1911) matriculated Christ Church 1869, B.A. 1873, M.A. 1876, Regius Professor of Pastoral Theology and a Canon of Christ Church 1885, became Dean of Christ Church, Oxford, in January of 1892, following the retirement of Henry George Liddell.

of open day—to kindly light. And thus was it wheeled down the High-street, over the bridge which crosses the sluggish Wey, and thence slowly climbed up the steep, stern slope to the gates of what must be one of the most lovely burying grounds in all England.[1]

The grave stands almost on the very crest of the ridge, which commands a glorious panorama of some of the finest scenery in the South of England. Deep down below you, as in a cup, lie the red roofs of Guildford, the winding river, the crumbling remains of the ancient Castle, beneath whose ivy-covered walls nestles "The Chestnuts," where the Dodgsons live. It is almost in a bee line with us, and even at this distance it is easy to pick it out from amongst its neighbours by the white blinds still down. Away beyond lies London; in the misty middle distance, on the sombre pine-clad heights, looms the Church of St. Martha's, where once Canterbury pilgrims lingered; to the right, the downs stretch as far as the eye can see. It is a grey day, and dull enough, with a wintry wind whistling weirdly amongst the churchyard shrubs, when views and prospects are least appreciated; and yet one lingers by this grave of "Lewis Carroll's" long after priests and surpliced choir have wended their way down the hill again. How lovely must the scene be in the summer sunshine; or bathed in the light of the harvest moon.

But Dean Paget is committing the dead to the earth: "Earth to earth, ashes to ashes, dust to dust"—it is not earth, or ashes, or dust, but chalk, all lumps of hard chalk, which fall upon the coffin. Verily such a spot does rob death of half his terrors—the worms, the damp, dank earth, full of creeping things. But the chalk has given forth sufficient sustenance to nourish a pine of goodly stature. Its sombre boughs, swaying to and fro in the wind, hang right over the open grave, and shelter the bare-headed clergy from its shrewd nipping. The sweet voices of the choir boys now rise:

"Peace, perfect Peace, in this dark world of sin,
The blood of Jesus whispers peace within—"

Words and music mingle with the winds, and are quickly lost to the ear; final prayers, too—and so the service ends, and all that remains of the famous writer is alone on the crest of this Surrey hill. Like Stevenson in the wild Samoan forest,[2] he rests on high.

The grave is quickly filled with the chalk; the surface is hid beneath the flowers. Many wreaths came from Oxford; some from children.[3] In

---

[1] Dodgson is buried at the cemetery near the top of The Mount, Guildford.

[2] Robert Louis Stevenson is buried on the island of Samoa where he died in 1894.

[3] For a complete footnoted list of wreaths and senders, see below, pp. 158–9.

particular we notice one or two, the inscriptions on which were in the faltering round hand of children—"From two of his child friends"; Enid and Margaret; Mary and Cuckoo; "to the sweetest soul that ever looked with human eyes"—and so forth. Such are few in number, but they must represent the feelings of innumerable children who loved Alice, who have visited the strange, topsy-turvy world known as "Wonderland" which "Lewis Carroll" peopled with his fanciful creations.

The Misses Dodgson, the Rev. S. H. Dodgson, Mr. W. L. Dodgson, Mr. C. H. Dodgson, Mr. J. H. D. Dodgson, the Rev. T. J. Prout, the Rev. T. B. Strong, Mr. A. G. V. Harcourt, Mr. G. L. Craik, Mr. Hassall, and one or two others may be named as being present at the ceremony.[1]

It is a remarkable fact, by the way, that "Lewis Carroll" is the first of eleven brothers and sisters to have died. He was sixty-six; the united ages of the eleven was seven hundred and odd years.

It is yet too early to say whether any biography of the dead man will appear,[2] but I believe I am correct in saying that he has left a large number of manuscripts behind him; and it is said that he has kept all letters which have been addressed to him for over forty years.[3]

Manchester Guardian
January, 1898

Everyone will be sorry to hear of the death of "Lewis Carroll," which we record this morning. He had endeared himself to successive generations of children for thirty years past, and their elders could not but admire him too. There was something delightfully paradoxical about the man, as well as about his writings. Mathematicians are seldom given to jesting, yet the Rev. C. L. Dodgson, the creator of the Jabberwock, the Hatter, the March Hare, and many another whimsical personage, was a serious mathematical tutor[4] for a quarter of a century. Equally curious in its way was his persistent refusal to admit that he wrote anything but treatises on the Fifth Book of Euclid or other severely mathematical topics. Once, when asked to allow some of the verses in his books to be

---

[1] For a more complete footnoted list of mourners see below pp.157–8.

[2] The first biography, by his nephew Stuart Dodgson Collingwood, appeared by the end of the year (see pp. xviii–xix). Since then more than a dozen others have followed.

[3] A vast quantity of papers from Dodgson's rooms at Christ Church were burned following his death (see pp. xiii–xiv). While some letters to him survive, most that were in his possession at the time of his death were probably destroyed.

[4] Dodgson's post was actually that of Mathematical Lecturer.

incorporated in a collection of comic verse, he replied gravely that there must be some mistake—that his only published works dealt with mathematics and were scarcely suitable for the purpose indicated. It was remarkable, too, that the man who entered readily into the pursuits of his child-friends was by nature of a very retiring and even melancholy disposition. Few authors of our time have held themselves so studiously aloof from their fellow-men. For all that, "Lewis Carroll" was universally esteemed, and with good reason. His stories, a delightful blend of prose and verse, of the real and the fanciful, and—in his later books, especially *Sylvie and Bruno*—of the grave and the gay, are the work of one who was gifted with an exceptional insight into child nature. Both by young readers who are fascinated by the queer adventures and by their seniors who appreciate the genial drollery and half-veiled satire, *Alice in Wonderland*, *Through the Looking-glass*, and *The Hunting of the Snark* will be treasured for a long time to come.

The Melbourne Age
January 17, 1898

The death is announced of the Rev. Charles Dodgson, better known by his pseudonym of "Lewis Carroll," author of *Alice's Adventures in Wonderland*, and many other works.

Unknown Paper (Melbourne, Australia)
February 26, 1898

It would be, perhaps, an exaggeration to say of the death of Lewis Carroll, the author of "Alice in Wonderland," what Dr. Johnson said of the death of Garrick,[1] that "it has eclipsed the gaiety of nations," but it has certainly removed from the stage the purest and most original humorist the English race has known since Charles Lamb. And Lewis Carroll—or the Rev. C. L. Dodgson, to give him his proper name—had a richer, quainter, and more pungent quality in his humour than even Lamb could boast. Perhaps, indeed, Lewis Carroll was himself a more entertaining paradox than any to be found in the long procession of amusing figures that waltz through the pages of *Alice in Wonderland* or the *Hunt-*

[1] David Garrick (1717–1779), actor and stage manager.

ing of the Snark. "Lewis Carroll" and C. L. Dodgson, as a matter of fact, represent two different men, and a tangle of personality as bewildering, if not quite as tragical, as that of Dr. Jekyll and Mr. Hyde.[1] Mr. Dodgson was a decorous divine, a hopeless and incorrigible bachelor, an absorbed mathematician, wandering dreamily through a world of tangents and curves and algebraic formulæ. "Lewis Carroll" was an unbuttoned humorist, the joy of all childhood, whose fancy took such wild flights and romped through such giddy freaks as no other humorist our somewhat sober and heavy-witted race has produced, has ever attempted. What human paradox, indeed, more puzzling can be imagined? The man who took a "first" in mathematics wrote the *Hunting of the Snark*! The shy and melancholy divine, who never got past deacon's orders because the pulpit was a place of terror to him,[2] was able by his mere presence to send a roomful of children into shrieks of laughter! This stammering, hopeless bachelor understood child nature as perhaps no mother ever did! He certainly held the key of a child's imagination, of its mirth, its fancy, its fears, in a degree that few mothers know. The man who in 1860 delighted mathematicians by publishing *A Syllabus of Plane Algebraic Geometry*, in 1865 added a new joy to English childhood by publishing *Alice in Wonderland*—a book which will fill with laughter the lips of generations yet unborn!

It would be difficult to exaggerate the quality of Lewis Carroll's humour. It had a riotous element in it which seemed to mock all restraints, and mounted to realms of the wildest burlesque. His daring and unrestrained humour, indeed, his eagerness to kindle at any cost the laughter of children, sometimes betrayed him into the most absurd situations. Thus the story is told that, "having been invited to a children's party, it occurred to him that a pleasant sensation might be produced if he entered the room on all fours. Unfortunately he mistook the number of the house, but did not discover his error until, warned by a strange silence instead of the peals of laughter he expected, he looked up from the floor, and was horrified to find himself in the midst of a party of adult strangers!" This extravagance of humour, indeed, runs through all Lewis Carroll's books, and perhaps makes them, for dull people in whom the sense of humour is nonexistent, unintelligible. The whimsical quality in his humour, the delicious muddle of his ideas, seem to suggest, as one critic says, that "one hemisphere of his brain had somehow got mixed with the other."

[1] *The Strange Case of Dr. Jekyll and Mr. Hyde* was published by Robert Louis Stevenson in 1886.

[2] In fact, Dodgson preached on many occasions—see for example, pp. 5n, 116n.

But through the dance-shadow of Lewis Carroll's absurdities there runs a clear thread of meaning, of a satire which is apparently unconscious and of milk-like sweetness, but which somehow has the sharpness of a sword-edge.

It is said that Mr. Dodgson was half-ashamed of the exploits of "Lewis Carroll." The mathematician and the divine, in brief, would sigh or blush over the excursions and exploits of the humorist. But it gives an enduring quality to Lewis Carroll's humour that it had as a background the strong brain of the mathematician and the grave conscience of the theologian. This explains, too, the circumstance, so rare in literature, that the same books which are the delight of childhood are the joy of grave politicians and of solemn philosophers. It would be absurd to say that Lewis Carroll's humour has a Shakespearian range but is there any other writer in English literature since Shakespeare who has found in the same degree the art of delighting both young and old, or who has supplied so many references and illustrations which have taken root in the common speech of the race? What is the greatest child's book in the English language to-day? It is not *Robinson Crusoe*,[1] nor *Gulliver*,[2] nor *Tom Brown's School-days*.[3] It is *Alice in Wonderland*. And yet, perhaps, the happiest and most effective metaphors in the speeches of politicians are culled from the same immortal volume. What could be happier, for example, than the use of *Alice in Wonderland* made by Mr. Symon, Q.C., in the Federal Convention only this week? The federal sentiment, he said, was fading away like the Cheshire cat in "Alice," until "soon nothing would remain of it but the grin"!

The world of to-day carries a heavier burden of care than the world of yesterday. It is perplexed with more serious problems; it is more strenuous and serious, less disposed to laughter, more apt to scream. There is some ground for the alarming suggestion that the very faculty of humour amongst us grows faint. No new Lamb arises. A second Dickens who could give us a new Pickwick or a new Sam Weller[4] translated into contemporary speech, is not so much as to be imagined. The choir of "new humorists" is a very distressful corporation indeed, and anything more dolorous than their jests could hardly be conceived. And yet how blessed

---

[1] *The Life and Strange and Surprising Adventures of Robinson Crusoe of York, Mariner* (1719) by Daniel Defoe (1616–1731).

[2] *Travels into Several Remote Nations of the World by Lemuel Gulliver* (1726) by Jonathan Swift (1667–1745).

[3] (1857) novel of school life at Rugby (the school attended by young C. L. Dodgson) by Thomas Hughes (1822–96).

[4] Sam Weller was a central character in Dickens' *The Pickwick Papers* (1836–37).

are the uses of humour! How wholesome is the office it discharges! Humour is the great antiseptic of life. It keeps thought sweet and clean. It has for the world of men the office salt has for the sea. If the sense of humour is indeed perishing from amongst us, then, alike for our politics, for our literature, and for our theology, as well as for our social life, the prospect is indeed black. But a generation which has produced and enjoyed Lewis Carroll is certainly rich in the elements of humour. Of *Alice in Wonderland* no less than 50 editions have been sold; copies of the first edition are already the joy and the prize of collectors.[1] These are circumstances which do honour to the generation. The truth is that humour, like every other intellectual quality, has some of the characteristics of a chameleon. It takes the colour of its environment. It talks the special dialect of its generation. It reflects the mood of its age.

The Montreal Daily Witness
January 18, 1898

The London papers are filled with long and affectionate notices of Lewis Carroll, whose death occurred the very day all England was set laughing by the reproductions in the *Times* of his famous trial scene of the Knave of Hearts, which weirdly fits the procedure in the Dreyfus case. With characteristic reserve no obituary mentions the well-known fact that he wished in his late years to marry Alice Liddell, the daughter of the Dean of Oxford, for whom the immortal fantasies were originally written, and that the decision of her parents that he was too much her senior broke two hearts.[2] He latterly rather disliked his humorous works, and deplored the fact that they obscured his real title to fame as a mathematician.

---

[1] In fact, the true first edition of *Alice* is one of the most famous rarities in all of book collecting. Displeased with the quality of the printing, Dodgson suppressed the edition and fewer than two dozen copies are known to have survived.

[2] Though Dodgson may have contemplated marriage with Alice when both were still young, and there has been much speculation that the break between Dodgson and the Liddell family that occurred in 1864 had to do with a suggestion of marriage on Dodgson's part, it is impossible that either could have desired marriage in Dodgson's "late years." Alice married Reginald Hargreaves in 1880.

The Morning January 15, 1898

With great regret we have to announce the death of the Rev. Charles L. Dodgson ("Lewis Carroll") at the Chestnuts, Guildford, the residence of his sisters, in his 66th year.

All children know "Lewis Carroll" as a household name. Generations have been brought up on *Alice in Wonderland* since its first publication just 32 years ago, and amid a host of rivil [*sic*] nursery tales and innumerable imitations which it has called into being, that masterpiece of the children's library has never been surpassed or equalled. To children of older growth it was left to appreciate its full fund of humour. Much of its fun is, no doubt, above the level of the little heads for which it was intended. But "Alice" in

Thousands Of Nurseries

has been a very real, very much alive young person, and the Griffon [*sic*], the Mock Turtle, the Mad Hatter, and the March Hare, and those others of that marvellous host by which Alice was surrounded in her adventures have retained an enduring place in memory long after childhood's days have passed, and in maturing middle-age have not lost their powers of delight.

And the nursery's hero—who was he? Some well-blessed paterfamilias, it might well be thought, with a host of children at his knee, to whom he retailed these marvellous tales round the fireside. On the contrary, Lewis Carroll was an elderly bachelor, a scholarly recluse, a mastermind in mathematics, as his many writings attest—in fact, everything that the creator of Alice might be expected not to be. He has himself told how the book grew. It was originally a little pamphlet, of some

Twenty Pages Or Less,

written in a neat, round hand, with crude but still humourous illustrations by the author himself, for the edification of a little relative to whom he was much attached.[1] In that favoured nursery "Alice" brought visions of new delight. Mr. Dodgson enlarged his book, and sought the wider audience of children that only its publication could secure. Sir John Tenniel—still happily among the living, at an advanced age—put some of his finest and most enduring work into the famous illustrations, from which Alice can never be dissociated, and *Alice in Wonderland* took the nursery world by storm.

---

[1] The original manuscript, *Alice's Adventures Underground*, written for Alice Liddell, who was not a relative, was ninety pages long. See above pp. 6n, 38n.

Through the Looking Glass followed in 1871. In this book, as fine in every line as its famous predecessor, Mr. Dodgson further developed that fund of melodious and nonsensical versification in which to this day he stands pre-eminent. It was the *Looking Glass* book which introduced us to "The Walrus and the Carpenter," so often quoted and parodied that its lines are among the best known

In English Verse,

to Tweedle-Dum and Tweedle-Dee, the Red and the White Queens, the Knave of Hearts,[1] and others whose names are familiar wherever English is spoken. With these two books alone "Lewis Carroll" was assured of a nitch [*sic*] in the temple of fame as a nursery classic, but they by no means represent his literary output on the lighter side. The *Hunting of the Snark* was written for the generation before this, while *Silvie and Bruno* reminds us that "Lewis Carroll" was still an active writer for the nursery within the past few years.

During his quiet life in Oxford Mr. Dodgson was a figure much sought after by visitors to the University city, but he was of a retiring disposition, and his intimacy was enjoyed by very few. Look up his name in a publisher's list, and you will find him as the author of an admirable *Elementary Treatise on Determinants*; the exponent of *Euclid, Book V., Proved Algebraically*; a teacher of *Symbolic Logic*—but not as the author of the books by which his name is immortalised. His authorship of *Alice in Wonderland* has, however, been an open secret for many years.

Mr. Dodgson was senior student at Christ Church, Oxford, and his association with the active life of the University remained unbroken until his death.

The Morning
January 15, 1898

To the present generation of children the death of Lewis Carroll means less than it would have meant to their predecessors. For some years past Mr. Dodgson had ceased to write, and his last book was by no means the equal of those classics, *Alice in Wonderland* and *Alice Through the Looking Glass*. To-day children devour those immortal works as eagerly as did their parents over a quarter of a century ago, but they had ceased to look upon their author as a man from whom more might be

[1] The Knave appears in *Alice*, not *Looking-Glass*.

expected. Mr. Dodgson was a type of the old-fashioned college Don, and devoted as he was to children it was said of him that he lost interest in them when they grew up, for he was a very shy man, and young ladies overpowered him as much as little girls attracted him. The story used to be told that the Queen, having read the *Alice* books with delight, sent to her bookseller for the rest of the author's works, and was astonished to receive the learned mathematical works which had made Mr. Dodgson's more serious reputation. Only yesterday a correspondent of the *Times* quoted a passage from *Alice in Wonderland*, which, absurd as it seemed at the time, was a perfect forecast of the procedure by which Captain Dreyfus was condemned. With the death of Lewis Carroll a very lovable personality passes away, and one which will be dear to children of all ages as long as the English language endures.

The Nassau Literary Magazine (Princeton University)[1] February, 1898

In Wonderland at close of day
We love—with you as guide—to stray
All open-eyed and fancy free
And children once again are we,
As with your elfish sprites we play.

It was a merry, happy way
You led us by. Yes, it was gay
And full of fun and jollity
    In Wonderland.

Within its borders would we stay
Whene'er the sky is dark or gray,
From tasks and serious things to flee
And follow your strange fantasy
With you, our guide, to be alway
    In Wonderland.

---

[1] This poem, titled "To Lewis Carroll," was signed "R. D. Dripps," and written by Robert Dunning Dripps, Princeton Class of 1898.

Nature (London) January 20, 1898

A formidable champion of Euclidean methods in the elementary teaching of geometry has just passed away after a short illness. The Rev. Charles Lutwidge Dodgson was born in 1832 at Daresbury in Cheshire; and, after passing five years at Rugby School, matriculated in 1850 at Christ Church, Oxford, where he was appointed a student in 1852, and graduated in 1854 with honours in both classics and mathematics. He was appointed Mathematical Lecturer in the College in 1855, and retained that office till 1881; he further served the University as Mathematical Examiner in 1863, and Moderator in 1868.

The mathematical subject in which he was most interested was the elementary teaching of geometry; of this he had a personal experience of twenty-six years. Without stint of labour he submitted to rigid logical analysis every text-book on the subject that came to his notice, undismayed by their surprising number, the result being the amusing and, at the same time, deep *Euclid and his Modern Rivals*, published in 1879, in which he demonstrated the logical superiority of Euclid's method over all the others examined. The Appendices of this book are very valuable. A "Supplement" to it appeared in 1885. In 1882 he edited Euclid, Books I. and II., with an introduction; and in 1888 he published *A New Theory of Parallels*, in the third edition of which (1890) he simplified his fundamental axiom.

His other mathematical work comprises *A Syllabus of Plane Analytical Geometry* (1860), *Formulae of Plain Trigonometry* (1861), *An Elementary Treatise on Determinants* (1867*), Euclid, Book V., proved Algebraically* (1874), and *Pillow Problems* (1893). He invented a new method of evaluating determinants, which is published in the *Proceedings* of the Royal Society for 1866, and also a method (which was published in *Nature*) of easily determining the day of the week corresponding to any date. In October last he described in *Nature* a brief method of dividing a given number by 9 or 11; and a second paper on the same subject, which appears in our correspondence columns this week, probably represents his latest contribution to mathematics.

A characteristic of all his work was the absolute exactness of expression at which he aimed, so that his definitions and proofs should be logically perfect. This carried a certain severity into his work, since, as he has himself remarked, a semi-colloquial style is apt to be also semi-logical, as nothing is more easy than to forget, in an argument which is interwoven with illustrative matter, what has, and what has not, been proved. It further tended to require the repetition of what might for exactness have to be a somewhat cumbrous periphrasis, to prevent which, therefore, he introduced a number of new terms and symbols; few of these have, however, been adopted into general use, though of the latter some are extremely expressive, and in his hands were of great value.

Mr. Dodgson's mind was essentially logical, in spite of the whimsical humour which has endeared "Lewis Carroll" to every boy and girl—nay, every adult—in the kingdom; and of late years he devoted a large part of his time to the study of the syllogistic methods of formal logic. In 1887 he published *The Game of Logic*, and in 1896 *Symbolic Logic, Part I.*

A shy and retiring man, he was to his friends a most charming companion, overflowing with the quaintest of humour, and one whose love for children was typical of himself, and whom to know was to love.

News of the Week
January 1898

Lewis Carroll died on the 14th of this month. I cannot write a column for boys and girls without mentioning the fact, although it may not be news to you. His right name was the Rev. Charles Lutwidge Dodgson. His body was interred last Wednesday at Guildford. An inscription, attached to a floral tribute composed entirely of arum lilies, was signed with the initials E. E. T. [*sic*], on it were written the following words:—"In loving memory of the sweetest soul that ever looked with human eyes." We can believe this as we read *Alice in Wonderland*, for surely no volume makes us long more to clasp the hand of its author than does this charmingly and beautifully told history of a child's dream. And now that hand is still! Mr. Dodgson was a very learned man. When Her Majesty, Queen Victoria, had read *Alice*, she expressed a wish to see the author's next work. And so a heavy treatise on the higher mathematics, if I remember rightly, was sent for her perusal.

New York Tribune[1]
January 16, 1898

The death of "Lewis Carroll" (the Rev. C. L. Dodgson), author of *Alice in Wonderland*, has caused the greatest regret in all parts of Great Britain. The papers are full of reminiscences, showing how in-

[1] This piece was evidently off the cable and sent to many papers. The same item or a

tense was his love for children and how universal was his shyness and dignity to others. "Lewis Carroll" was extremely particular as to his personal appearance. His alert figure and general vigor conveyed the impression that he was much younger than sixty-five. He never wore an overcoat in his life and continued to take long constitutionals to the end.[1] He was staying with his sisters at Guilford [*sic*] when he died.

New York Tribune
January 16, 1898

The comic genius has had none too many incarnations. From Aristophanes down, the masters have stood upon ground apart, forming a small and peculiarly distinguished company. Lewis Carroll was one of that choice assemblage. He was, in his extraordinarily original field, as much a master of the comic spirit as Moliere,[2] and his death leaves an irreparable gap in modern letters. This, no doubt, seems excessive praise to those who have read *Alice in Wonderland* in vain. It is true, we believe, that there are persons who resent the episode of the Walrus and the Carpenter as a violation of the laws of Nature; who cannot understand all this fuss about the Snark; who gravely ask you if a portmanteau word is to be found in the dictionaries. To them the gates of poetry are closed. They have never known the rapture of an uffish thought. A vorpal sword would be useless in their hands, and from a manxsome foe they would fly shrieking.

Lewis Carroll was emphatically a poet. We have no patience with the hypothesis that connects his wonderful ingenuities and conceits with his training as a mathematician. You cannot reduce the Cheshire cat that faded away until nothing but its smile was left upon the bough to any terms of literary artifice. That cat was a pure inspiration. All of Carroll's humorous productions were the fruits of a strong intellect, turning for relaxation to sublime nonsense as to the divine afflatus, and in the realm of that nonsense preserving its poise, its sobriety, its constructive power,

---

nearly identical item appeared in the following papers: *Los Angeles Herald* (16 Jan. 1898), *New York Herald* (17 Jan. 1898), *The Times*, Richmond Virginia (16 Jan. 1898), and *The Washington Post*, Washington, D.C.; and the mention of his not wearing an overcoat was included in other papers as well.

[1] Dodgson frequently recorded his walks in his diary, and in the last years of his life mentions several "regular" walks of seventeen miles or more.

[2] Jean Baptiste Poquelin (Moliere) French comic playwright (1622–1673).

with the instinct of a poet. He lays down the most grotesque propositions, in the most preposterous terms, with an air of perfect seriousness. In other words, his topsy-turvy world, like Mr. Gilbert's, is jealous of its laws, and acknowledges their sway with punctillious gravity.

This is one of the secrets of the truest wit, the most exquisite humor, this artless art of keeping one's self well in hand. The writer, the speaker, lends verisimillitude to his wildest extravaganza through appearing to believe in it himself as a matter of course. The listener, the reader, is excited by this seriousness of treatment as well as by the intrinsic fun of the thing narrated, and he is dissolved in laughter. Lewis Carroll held the keys of mirth on these terms, and held them puissantly if ever a man did. Perhaps the explanation of his great fame may be found not only in the rare note of his genius, but in the felicity with which he somehow contrived to appeal with equal force to young and old alike. Children adore his Alice, and so do men of mind. Bayard Taylor[1] used to stalk up and down pouring forth the Jabberwocky until his delight in the sheer poetry of the thing was whelmed in Homeric laughter. Happy Lewis Carroll! It is impossible to think of him, even over his grave, without a sense of sunshine and of joy. In the Elysian Fields the humorists take him by the hand, Shakespeare and Cervantes[2] greet him as a friend and comrade.

New York Tribune (Illustrated Supplement)[3] January 29, 1898

The devotee of Lewis Carroll is loath to put down the books that have been taken up again upon the occasion of his death. They are among the most beguiling works in English literature. In that long row of volumes published by the Macmillans one is lured on from one to the other with no consciousness of the passage of time. Yet the sky is not all untroubled. One or two of these productions awake faint regrets. Why, for example, did *Sylvie and Bruno* get itself written, and why—more pressing question—did *Sylvie and Bruno* get itself "Concluded," both of them in volumes far too long? In these Lewis Carroll failed to recapture his early magic. He is still Lewis Carroll, and the *Sylvie* books

---

[1] American poet and author (1825–78).

[2] Miguel de Cervantes Saavedra (1547–1616), Spanish novelist, dramatist, and poet, author of *Don Quixote*.

[3] This article was accompanied by a reproduction of Tenniel's illustration of the Jabberwock.

will be cherished affectionately. But their famous predecessors will be held in even more loving remembrance.

There is a great tenderness in the regard which Lewis Carroll aroused among his readers. For one thing, no humorist so utterly frantic as the author of "The Jabberwocky"[1] was ever so exquisitely refined. Part of the charm of *Through the Looking Glass* and *Alice's Adventures in Wonderland* comes from an atmosphere of innocence, of sweetness, which sends the imagination adrift among senses far away in the woods, where fairies dwell and where children come into their own. These books are peculiarly unworldly. Their fun provokes no laughter of the sort that is compared to the crackling of thorns under a pot. A kind of sub-current of fairy runs through every page, an echo of the elflike happiness of childhood. If any doubt remains of this depth that is so important in Lewis Carroll, let the reader turn to the serious lines prefixed to *Through the Looking Glass*:

> Child of the pure unclouded brow
> And dreaming eyes of wonder!
> Though time be fleet, and I and thou
> Are half a life asunder.
> Thy loving smile will surely hail
> The love-gift of a fairy tale.
>
> I have not seen thy sunny face,
> Nor heard thy silver laughter;
> Nor thought of me shall find a place
> In thy young life's hereafter—
> Enough that now thou wilt not fail
> To listen to my fairy tale.
>
> A tale begun in other days,
> When summer suns were glowing—
> A simple chime, that served for time
> The rhythm of our rowing—
> Whose echoes live in memory yet,
> Though envious years would say "forget."
>
> Come, hearken then, ere voice of dread,
> With bitter tidings laden,

[1] "Jabberwocky."

Shall summon to unwelcome bed
A melancholy maiden!
We are but older children, dear,
Who fret to find our bedtime near.

Without, the frost, the blinding snow,
The storm-wind's moody madness—
Within, the firelight's ruddy glow,
And childhood's nest of gladness
The magic words shall hold thee fast;
Thou shalt not heed the raving blast.

And though the shadow of a sigh
May tremble through the story,
For happy summer days gone by,
vanish'd summer glory—
It shall not touch, with breadth of bale
The pleasance[1] of our fairy-tale.

The man who could write this could also write this other immortal stanza:

And as in uffish thought he stood,
The Jabberwock, with eyes of flame,
Came whiffling through the tulgey wood
And burbled as it came![2]

There seems to be a great gulf between these two notes. But they were notes of the same endearing voice; they sprang from the same fine imagination, and the mark of that imagination is really upon them both. In fact, Lewis Carroll had that indescribable cachet of genius which makes every work of an author characteristic, no matter how different it may be from the one that preceded it. The Macmillans publish in facsimile the original manuscript of *Alice's Adventures Under Ground*, from which the *Adventures in Wonderland* were developed. It is the true forerunner of the author's more famous book, it is full of his peculiar charm. *The Hunting of the Snark* and all those other pieces which accompany it, *Phantasmagoria*, "The Three Voices," "Hiawatha's Photographing" and

[1] Alice Liddell's middle name was Pleasance.
[2] From "Jabberwocky," in *Looking-Glass.*

the shorter productions, are less glorious, perhaps; than "The Jabberwocky," but they have the same captivating absurdity and the same poetic fire which we recognize in the latter masterpiece. And even to his mathematical books the author of *Alice* was true to himself. *Symbolic Logic* is full of problems for austere minds, but the terms in which the puzzles are expressed are irresistibly droll. *The Game of Logic* subjects the student to a less engaging, if less exacting, discipline, but in *A Tangled Tale* some really abstruse and fascinating mathematical problems are handled with almost screaming humor. *Curiosa Mathematica* and *Quarternions*[1] we do not know, but we venture to assume that they make hard study amusing, if nothing else. It is a strange bibliography. It is amazing to think of the same man producing *Through the Looking Glass* and *Symbolic Logic*. But, we repeat, an acquaintance with these books soon reveals their kinship.

It is very much to be hoped that all of Lewis Carroll's posthumous manuscripts that he would have cared to print will see the light, and that some memoir of him will be published. He must have been a charming companion, and he must have written delightful letters.[2] He led a quiet and a secluded life. We do not know that his portrait has ever been published. He never wanted publicity, and perhaps it would be kinder to leave him to his privacy now. Yet there must be thousands who long, in all reverence and delicacy, to know something of the personality of the man to whom they have owed so much. The person whom we would choose to portray Lewis Carroll would be that Alice Liddell for whom, as it has long been understood, he wrote his *Alice* books. It is something of a coincidence, by the way, that her father, Dean Liddell, should have died only a few days after Lewis Carroll.

New York Tribune
February 2, 1898

The latest numbers of the English literary weeklies reaching these shores devote more or less space to the late Charles Lutwidge Dodgson, otherwise Lewis Carroll. Though he was, of all recluse souls,

---

[1] Not a work by Dodgson. Perhaps a reference to *Lectures on Quatermions* by Sir William Rowan Hamilton (1805–65) or Peter Guthrie Tait's (1831–1901) work on quatermions.

[2] Dodgson wrote some tens of thousands of letters during his life. Many are published in Cohen, *The Letters of Lewis Carroll.*

the shyest, it is safe to say that the author of the *Alice* books would not have winced under appreciation generally so sympathetic, even affectionate. There is an unwonted gentleness in the comments of that usually acrid sheet, *The Saturday Review*.[1] More than any other English writer it declares Carroll has attracted to himself the crabbed age and the youth of his generation.

For that, mere cleverness cannot account: to have taken such root, to have become so naturalized, his work must have struck a note of peculiar appropriateness to his day; it was as it were a microbe that fell upon a constitution happily prepared for its inroads. Yet his rendering of fairy tale was a departure from tradition.

Of all true fairy tale this may be said: It has never been "written down" to the intelligence of childish readers; rather, it has been the meeting-point of interests, the medium through which the child can experience the thrill of big-hearted adventure, and the grown man the romantic sense and undisturbed belief in poetic justice which are the birthright of childhood. For centuries the child has warmed its imagination upon the folk tales which once gave contentment to its less simple-minded elders, and do still have an interest for them more or less vague. But up to the latter half of this century the workings of a child's mind were of little interest to its elders. Now, however, an astonishing change has come over the dream of mankind; and to-day the child is the grown-up man's fairy-book. Almost by instinct a wise mind stoops to look into that under world so close at hand, so stored perhaps, with the wisdom of the ancients; and so a new ground for fairyland has been reclaimed from the prose of existence. To minds still sensitive to chivalry and romance, the old fairy tale remains true; but a far larger world of believers has been gathered to the imaginative world of childhood and arrested, almost studiously, to a contemplation of its laws.

It was through this change of outlook, and as almost its first portent, that Lewis Carroll's fairy books struck the light. Without condescension, without any "putting back of bright intelligence," rather with all the resources of wit and humor, and a sub-pathetic sense of the ridiculous, Carroll, a man of grave reserve and fine intelligence, turned round upon all tradition to find a new fairyland ready-made, fresh-sprung from humanity's new interest in the things of childhood.

The seriousness of his work and its safety come from the fact that it is as much fairy tale for elders as for children; and the highest interest of

---

[1] See below pp. 139–42 for the complete, footnoted text of this article.

these two time-divided classes is the interest they have in common. Lewis Carroll's ability to evoke this common interest was the measure of his success, the material for it was his great discovery.

It may seem strained, in the face of the mirth-moving spirit of the *Wonderland* and *Looking-glass* tales, to claim for them so much purpose and seriousness; but few can feel that their quaint wit and baffling logic were but the outcome of a mere sense of fun. Wherever there is beauty in human things there must also be pathos, something intent if not wistful; and in many of Carroll's tender extravagances there is beauty of a high order. Take, as an instance, the Fawn that has forgotten its name and accompanies Alice fearlessly till the wood is cleared; then, crying, "I am a fawn, and you—you are a human child!" springs startled away. So, also, the whole episode of the White Knight, of so much wisdom gone foolishly astray, has about it almost the same charm as Millais' "Sir Isumbras," now delighting all who make Burlington House their mild wintry resort. There is satire, too, light and stingless, but satire none the less, conveyed under a cajoling disguise of parody; and some may be able to see in the "aged, aged man a-sitting on a gate" a not distant reference to Wordsworth's "leach-gatherer," and in the figure of the White Knight, with his horse-like face, some likeness to the poet himself.

Lewis Carroll was too much of a scholar and a recluse in temperament not to find many things contrary to his taste in the world from which, almost with ascerbity, he guarded his private personality; its slipshod speech and illogical ways of thought received many a passing rap from his wit; and the desire for interrogation and argument which his heroine Alice, charming type of her age, carries to the length of foolishness, is resolutely put down with a finality which leaves the incorrigible little offender breathless. . . It seems as if Lewis Carroll was sent to poke fun with unexampled urbanity at the new age which has sprung upon us and yet, from many grave and tender reasons, to welcome it.

The Saturday elsewhere sets forth a fact which may be commended to the so-called "literatures" of the period.

In these days of personal puffs and self-advertisement the career of the late Lewis Carroll is a trenchant object lesson. By a single inspiration of unconscious genius, he achieved an international fame which others, as great as he, only partly win through travail and despair. It is no hyperbole to say that there is not a spot in the civilized world, not a language with any pretensions to a literature, to which the Jabberwock and the Cheshire cat are unknown. Yet what did the great reading public know of the quiet student of Christ Church? What significance did the name of Charles Dodgson bear to the average mind? About the little humorists of the hour, their wives, their clothes, their pastimes, we are told in every evening newspaper, but the king of drolls was content to speak only through his works, and their voice is more lasting than all the trumpets of the noble army of log-rollers put together.

The Academy[1] notes that Lewis Carroll's attitude to children was more devotion than mere affection, approaching even to adoration. He wrote *Alice in Wonderland* to give pleasure to two friends, the little daughters of Dean Liddell, the elder of whom was the original Alice.[2] It was published that others might share that pleasure. It was his wish that the Lewis Carroll of his books should be kept apart from the Rev. Charles Dodgson of his private life, and *The Academy* mentions a proof of this desire, which is to be found in the letter he wrote to Mr. R. H. Caine, the editor of a collection of humorous verse, who asked him for permission to reprint some of his pieces in that volume.

Mr. Caine received this reply:

> Mr. C. L. Dodgson begs to say, in reply to Mr. Caine's letter received this morning, that he had never put his name to any such pieces as are named by Mr. Caine. His published writings are exclusively mathematical, and would not be suitable for such a volume as Mr. Caine proposes to edit.

Against this rebuff might be placed the following letter to a child (written in 1875), wherein the gulf existing between the two personalities is at once emphasized and removed; but it must be remembered that Mr. Dodgson would do for a child what he would not do for anyone else:

> My Dear Magdalen: I want to explain to you why I did not call yesterday. I was sorry to miss you, but you see, I had so many conversations on the way. I tried to explain to the people in the street that I was going to see you, but they wouldn't listen; they said they were in a hurry, which was rude. At last I met a wheelbarrow that I thought would attend to me but I couldn't make out what was in it. I saw some features at first. Then I looked through a telescope and found it was a countenance; then I looked through a microscope and found it was a face! I thought

---

[1] See above pp. 3–17 for the complete, footnoted text of items in *The Academy* which are quoted below.

[2] At the time of the composition of *Alice in Wonderland*, Alice Liddell was the middle of three sisters. Another sister, Rhoda, was born later.

> it was rather like me, so I fetched a large looking-glass to make sure, and then, to my great joy, I found it was Me. We shook hands and were just beginning to talk when Myself came up and joined us, and we had quite a pleasant conversation. I said, 'Do you remember when we all met at Sandown?' And Myself said, 'It was very jolly there; there was a child called Magdalen,' and Me said, 'I used to like her a little. Not much, you know—only a little.' Then it was time for us to go to the train—and who do you think came to the station to see us off? You would never guess, so I must tell you. They were two very dear friends of mine, who happened to be here just now, and beg to be allowed to sign this letter as your affectionate friends, Lewis Carroll and C. L. Dodgson.

His mathematical works were numerous and valuable, although his championship of Euclid against more modern systems of geometry is held by many to be fantastic. Mr. Dodgson had many of the eccentricites which so often accompany proficiency in his particular science and many good stories are told of him at Oxford. He was a watchful guardian of Oxford's honor, and used occasionally to put forth a whimsical pamphlet in which some phase of the University's wellbeing was examined. These productions were always witty and marvelously ingenious. Mr. Dodgson was shy and reserved, a resolute celibate, a man of few friends, but fit; and the patron saint of children. Incidentally we might mention that he liked them all to be familiar with Lewis Carroll's writings. His hobbies, after mathematics, which he looked upon both as work and play, were photography and the stage. His photographs of children must be well nigh countless. Mr. Dodgson—as sage, as wit and as saint—will be mourned by those who knew him as Lewis Carroll will be mourned by readers all the world over.

A correspondent of *The Academy* says that few of Oxford's famous men have been so inconspicuous in her midst. "Froude was constantly to be observed; even Walter Pater was known by sight to a large part of the undergraduate world, but I scarcely remember to have seen 'Lewis Carroll' half a dozen times in the street."

In Oxford his works enjoyed a surprising popularity, and formed the storehouse for undergraduate nicknames. In my own day it even became the fashion for a man to set them in foolish paradox by the side of Shakespeare, when incautiously questioned on his preferences in letters. *The Hunting of the Snark* was popularly supposed to contain all the metaphysics in the world. I once heard a distinguished college authority explain his course of lighter reading during one vacation. "The first week," he said, "I read *Sylvie and Bruno*. "And then?" some one asked. "And then," he said, "I read the Second Part." "And then?" "And then," he murmured in doubt "then," brightening up, "ah, then I went back to *Through the Looking-Glass*!"

Another correspondent recalls the fact that *Alice* was not on its first appearance received with much enthusiasm. *The Times* praised Tenniel's drawings highly, but remarked of the text only that it was "an excellent piece of nonsense." *The Athenaeum* gave the book a slap after this fashion:

> This is a dream-story but who can, in cold blood, manufacture a dream, with all its loops and ties, and loose threads, and entanglements, and inconsistencies, and passages which lead to nothing, at the end of which Sleep's most diligent pilgrim never arrives. Mr. Carroll has labored hard to heap together strange adventures and heterogeneous combinations, and we acknowledge the hard labor. Mr. Tenniel again is square and grim and uncouth in his illustrations, howbeit clever, even sometimes to the verge of grandeur, as is the artist's habit. We fancy that any real child might be more puzzled than enchanted by this stiff, overwrought story.

Strange to say, *The Athenaeum* still maintains a Philistine attitude. It publishes a very short and very cool obituary notice,[1] apparently feeling the necessity of supporting a character for consistency at the expense of its reputation for critical acumen. The most interesting thing in the short notice is its statement of what most people have forgotten—that "Lewis Carroll" is an anagram on Mr. Dodgson's Christian names Latinized, "Carolus Ludovicus."

The Literary World devotes some of its space to reminiscences by an old Oxonian, who once met Carroll at dinner at Christ Church.[2]

> The only Alice-in-Wonderlandy thing I remember hear [*sic*] him say was his wondering why a cat had only one tail. He was credited with having disposed of the Homeric problem by the solution that the *Iliad* and the *Odyssey* were not really written by

---

[1] See above pp. 18–19 for the complete, footnoted text of this article

[2] *The Literary World*, Boston, carried an obituary of Dodgson on 19 Feb. 1898, which was drawn from accounts in the British press. The editors have been unable to trace the article alluded to here (apparently of British origin).

> Homer, but by another man of the same name. He was a mathematical Don, and a good many years ago, when the unco' orthodox in Oxford thought that Jowett ought to be got out of the place, brought out an elaborate problem in which the thing to be done was "to eliminate J."[1] He once printed a fanciful description of a museum which was about to be constructed in Oxford. Besides other accommodation, it was to contain a large room for the greatest common measure, a small room for the least common multiple, a long passage in which, in the beautiful language of Euclid, lines could be produced ever so far, and an apartment for fractions reduced to their lowest terms—and undergraduates were to be allowed to keep their terms there, too.[2]

Those who knew Lewis Carroll best knew, as one of his friends has said, how behind all his varied activities lay a deep background of religion—"a religion severely quiet and retiring, like his character." Something of this feeling shines through the sweet and manly words he addressed as "An Easter Greeting" in 1896, to "Every Child That Loves Alice":

> If I have written anything to add to those stories of innocent and healthy amusement that are laid up in books for the children I love so well, it is surely something I may hope to look back upon without shame and sorrow (as how much of life must then be recalled!) when my turn comes to walk through the valley of shadows.
>
> I do not believe God means us to divide life into two halves—to wear a grave face on Sunday and to think it out of place to even so much as mention Him on a weekday. Do you think He cares to see only kneeling figures, and to hear only tones of prayer, and that He does not also love to see the lambs leaping in the sunlight, and to hear the merry voices of the children as they roll among the hay? Surely their innocent laughter is as sweet in His ears as the grandest anthem that ever rolled up from the "dim religious light" of some solemn cathedral.

---

[1] *The New Method of Evaluation as Applied to π* (1865), was one of Dodgson's pamphlets relating to the debate over the endowment of the Greek Professorship held by Benjamin Jowett (see Wakeling, *The Oxford Pamphlets, Leaflets, and Circulars of Charles Lutwidge Dodgson*).

[2] A paraphrase of an excerpt from *The Offer of the Clarendon Trustees* (1868). See below, p. 118, for a more accurate quote from this pamphlet.

New York Tribune
February 22, 1898

Speaking of the late Lewis Carroll, Lionel Tellemache[1] [*sic*] says: "The following specimen was propounded by him in my presence: Suppose that I toss up a coin on the condition that, if I throw heads once, I am to receive 1d.; if twice in succession, 2d.; if thrice, 4d.; and so on, doubling for each successive toss; what is the value of my prospects? The amazing reply is that it amounts to infinity: for, as the profit attached to each successful toss increases in exact proportion as the chance of success diminishes, the value (so to say) of each toss will be identical, being, in fact, 1/2d.; so that the value of an infinite number of tosses is an infinite number of half-pence. Yet, in fact, would any one give me sixpence for my prospect? This, concluded Dodgson, shows how far our conduct is from being determined by logic. The only comment that I will offer on this astounding paradox is that, in order to bring out his result, we must suppose a somewhat monotonous eternity to be consumed in the tossing process."

The New York Times
January 16, 1898

The Rev. C. L. Dodgson, whose nom de plume was Lewis Carroll, the author of *Alice in Wonderland*, is dead.

Mr. Dodgson was born in England in 1832 and was graduated from Christ Church, Oxford, as first-class mathematician in 1854. He published *The Adventures of Alice in Wonderland* in 1865, having remained at Oxford as lecturer on mathematics, tutor, and student, and except for his books, his life was as uneventful as that of any man in the university. In 1869 he published *Phantasmagoria*, a collection of humorous poems and parodies, and in 1871 children whose earliest recollections were of the delights of *Alice in Wonderland* were further charmed by *Through the Looking Glass and What Alice Found There*. *The Hunting of the Snark, an Agony in Eight Fits*, which has become almost as famous and familiar in the larger world as Alice is in the nursery, was published in 1876.

Mr. Dodgson's life was as grotesque in its contradictions as his most deliciously absurd conceptions. He was a profound student of math-

[1] Lionel Tollemache (1838–1919) knew Dodgson and wrote "Reminiscences of Lewis Carroll" in *Literature* (5 Feb. 1898), reprinted in Cohen, *Interviews and Recollections*.

ematics and wrote ponderous books on that exact science, turning aside to put into orderly words and fascinating form the chaotic, queer ideas and dreams of a child's unformed and untrained mind, and to weave nonsense words and the wildest of imaginary things and events into irresistible humor.

In addition to his better known works he wrote such books as *A Syllabus of Plane Algebraical Geometry*, *An Elementary Treatise on Determinants*, *A New Theory of Parallels*, and so on. He was also the author of *A Tangled Tale*, *Alice's Adventures Underground*, *Rhyme? and Reason? Doublets, a Word Puzzle*, and *Euclid and His Modern Rivals*.

He had no children of his own, having never married, but was devoted to children as they were to him, spending all his spare time with them and throwing aside his dignity when in their company. It is told of him that a company of dignified people at a London dinner table was once astonished by seeing a strange gentleman, evidently a clergyman, Mr. Dodgson having been ordained in 1861, enter the dining room on all fours. When matters were explained, it was found that the author of "Alice," having been invited to a children's party at a house near by, had mistaken the place, and made before the wrong company the entrance with which he had intended to amuse his young friends. As the children grew to manhood and womanhood, he lost interest in them, and turned from them to the next generation.

The original of Alice, whose wonderful adventures have probably delighted more children than any other book that was ever written, is said to have been Alice Liddell, daughter of a former Dean of Oxford, who afterward became Mrs. Reginald Hargreaves. She and her sister Edith were Mr. Dodgson's favorite playmates in their day, and his first idea of *Alice in Wonderland* is said by tradition to have originated one drowsy Summer day as he lay in the bottom of a boat while the two little girls were zealously splashing and tugging with the oars along the shining and tranquil surface of the river.

His children's books were illustrated by Tenniel, and Henry Holiday drew the curious pictures for *The Hunting of the Snark*.

The Outlook (London)
January 22, 1898

The Rev. C. H. Dodgson, better known as "Lewis Carroll," whose death occurred last week, was a unique genius. His specialty was mathematics, and it is said that latterly he rather disliked his humorous works, deploring the fact that they obscured his real title to fame as a mathematician. Unless we are greatly mistaken, however, his fame has been assured far more by *Alice in Wonderland* than by any of his treatises on mathematics. The grotesque fantasies, we are told, were originally written for Miss Alice Liddell, a daughter of the famous Dean. Mr. Dodgson's mathematical training stood him in good stead when he attempted the marvelous conceits which set two hemispheres laughing. Many have believed that such ingenuities were the logical "other side" of a mathematician's life—his natural, restful play. Be this as it may, the works of "Lewis Carroll" show to their admirers more than this. The world will remain convinced that, beneath the pseudonym which has long seemed a real name, there was the soul of a poet of humor, belonging to the great school whose representatives in many ages and countries have been Aesop, Aristophanes, Cervantes, La Fontaine,[1] Goldoni,[2] Moliere, Scheffel.[3] Such men have shown how nonsense serves sense. Jabberwocky talk has passed into the currency of the English language. Long may it remain there, each word a synonym of the mind-relaxation necessary to every sane human being. Surely the genius which gravely gave forth delightfully preposterous scenes expressed in a new and deliciously vivid language has created its own place in literature.

Oxford Chronicle[4]
January 22, 1898

We regret to announce the death of the Rev. Charles Lutwidge Dodgson, M.A., known to so many thousands of readers, old and young, as "Lewis Carroll." Mr. Dodgson was staying with his sisters at their residence at Guildford, where he died on Friday in his 66th year. Mr. Dodgson and Lewis Carroll were in authorship, and to some extent in life also, two very different beings. Mr. Dodgson was the don, and writer on mathematical subjects; Lewis Carroll was the author of charming books for children. Mr. Dodgson graduated at Christ Church, Oxford, in 1854—having taken a first-class in mathematics, and a third in Lit. Hum. In the following year he was appointed Mathematical Lec-

---

[1] Jean de La Fontaine (ca. 1621–1695), French fabulist.

[2] Carlo Goldoni (1707–1793), Italian comic playwright.

[3] Joseph Victor von Scheffel (1826–86), German comic poet and novelist.

[4] Portions of this obituary also appeared in *The Oxford Times* (22 Jan. 1898) and *The Westminster Gazette* (15 Jan. 1898). See below, pp. 120–22, 162–71.

turer to the college, a post which he continued to occupy until 1881. In 1858 he was elected to a Senior Studentship, and this he continued to hold until the end. In 1861 he was ordained. His earliest, as well as his latest, publications were all mathematical—*A Syllabus of Plane Algebraic Geometry* (1860), *The Formulæ of Plane Trigonometry* (1861), *A Guide to the Mathematical Student* (1864). The first of his children's books, to which we shall presently recur, was published in 1865. Other mathematical works followed, of which the more important were—*An Elementary Treatise on Determinants* (1867*), Euclid, book V., proved Algebraically* (1874), and *Curiosa Mathematica: A New Theory of Parallels* (1888). Considerable, however, as were Mr. Dodgson's contributions to mathematical science, and constant though his devotion was to mathematical teaching, his reputation in these respects was altogether overshadowed by the fame of Lewis Carroll. *Alice's Adventures in Wonderland* was originally written in MS. for the amusement of his girl friends at Oxford, and a facsimile of it in that form was published many years afterwards. It first saw the light in 1865, and at once became a popular favourite. The sequel *Through the looking-glass and what Alice found there*, published in 1871, was equally successful, and it has run through as many editions as the original *Alice.* Mr. Dodgson used to be fond of repeating the deliciously naive remark of a child-friend, whom he asked, after an acquaintance of two or three days, if she had read *Alice* and the *Looking-Glass.* "Oh, yes," she replied readily, "I've read both of them. And I think" (this more slowly and thoughtfully)— "I think *Through the Looking-Glass* is more stupid than *Alice's Adventures.* Don't you think so?" Next to the *Alice* volumes, Lewis Carroll made his principal hit with *The Hunting of the Snark: An agony, in eight fits.* This appeared in 1876, and had to be reprinted several times in the same year. It was subsequently embodied in a larger volume called *Rhyme and Reason* (1883). Of his remaining books, *Sylvie and Bruno* (1889), with its *Conclusion* (1893), is the best known. *Alice's* and *The Hunting of the Snark* have by this time become household words. Thousands of children have found infinite delight in them, and elder folk have found infinite applications for the author's wit and wisdom. Mr. Dodgson, as might be gathered from Lewis Carroll's books, was a genuine lover of children. There was no idea whatever of publication in his mind when he first wrote *Alice*; the verses which he afterwards prefixed to the book describe real scenes on real river expeditions with real little girls—the daughters of Dean Liddell. There are, indeed, if the asservations of all his girl friends are to be believed, innumerable "real Alice's." Mr. Dodgson's beautiful suite of rooms in the north-west corner of Wolsey's great quadrangle, looking over St. Aldate's, were at one time a veritable children's paradise. Never did rooms contain such endless stores of fascinating things. These rooms in Christ Church were a real Wonderland for many a real child. From Mr. Dodgson's cupboards musical boxes, mechanical performing bears, picture-books innumerable, toys of every description, came forth in bewildering abundance before the child's astonished eyes; no wonder, then, that in childish years a day spent with "Lewis Carroll" was like a glimpse into a veritable El Dorado of innocent delights. Nor were Mr. Dodgson's little friends all of one class. He took for many years a deep interest in homes and hospitals for sick children, and no inconsiderable portion of the profits of his books was devoted to these works of charity. Mr. Dodgson had beneath his twinkling humour a deep fund of seriousness and religious earnestness. He used at one time to issue a fly-leaf to "every child who loves Alice," in which he sought to mingle grave thoughts of charity and faith with fun and frolic. For many years Mr. Dodgson was a considerable amateur photographer, and amused himself by taking his little friends in all sorts of odd and fanciful costumes, till his albums became filled with Japanese boys and girls, beggar-maids in picturesque tatters, or Joans of Arc in glittering armour. The smell of the collodion he used to pour on to the negative, while his small "subjects" watched him open-mouthed the while, lingers in the memory of many whose childish days he helped to brighten. His photographic hobby appears in the volume called *Phantasmagoria.*[1] Mr. Dodgson was also fond of the theatre, and nothing gave him more pleasure than to take a party of his little friends to the play. It is not easy to realise that Mr. Dodgson was 65. His hair was indeed iron grey; but his alert figure and his bodily vigour seemed to betoken a younger man. He never wore an overcoat in his life, and he continued his long "constitutionals" to the end. Though somewhat of a recluse, Mr. Dodgson had many friends, both in Oxford and elsewhere, by whom he will be sincerely mourned, while the death of Lewis Carroll will be felt almost as a personal sorrow in nurseries and schoolrooms wherever English literature is known.

Among the best known of Lewis Carroll's jeux d'esprit issued at Oxford was "a monograph" on *The New Belfry of Christ Church, Oxford* (1872), to which strong objection was taken on artistic grounds. The monograph began as follows:—

1. On the etymological significance of the new belfry.

The word "Belfry" is derived from the French "bel"—beautiful, becoming meet—and from the German "frei"—free, unfettered, secure, safe.

---

[1] Presumably a reference to Dodgson's comic poem "Hiawatha's Photographing."

Thus the word is strictly equivalent to "meat-safe," to which the new belfry bears a resemblance so perfect as almost to amount to coincidence.

Some of Lewis Carroll's parodies of the propositions of Euclid were decidedly happy—as, for instance, "A discussion may be raised on any point at any distance from that point," and "A plain, superficial speaker is one who, any two points being given, will lie evenly between those points."

A good Oxford repartee has been attributed to several people, and among others to Mr. Dodgson. On one of his daily constitutionals he met an undergraduate driving a tandem. This undergraduate had a day or two before been driving one of the Egyptian princes then at Oxford, and had spilled His Highness. "May I give you a lift," asked the undergraduate as he passed Mr. Dodgson. "Wilt thou kill me," was the answer, "as thou didst the Egyptian yesterday?"

Sir,—In your interesting account of the life of "Lewis Carroll" you say that the original of Alice "has long been dead." This is not so. She married a Mr. Hargreaves, and now lives in the New Forest. It was her sister who met with an "untimely death." With regard to the remark that "Mr. Dodgson and Lewis Carroll were very different beings," the following letter, written to a child of eight, accompanied by a copy of *The Hunting of the Snark*,[1] is interesting:—

> Ch. Ch., Oxford, Dec. 15, 1875.
>
> My Dear Magdalen,—I want to explain to you why I did not call yesterday. I was sorry to miss you, but you see I had so many conversations on the way. I tried to explain to the people in the street that I was going to see you, but they wouldn't listen; they said they were in a hurry, which was rude. At last I met a wheelbarrow that I thought would attend to me, but I couldn't make out what was in it. I saw some features at first. Then I looked through a telescope and found it was a countenance; then I looked through a microscope and found it was a face! I thought it was rather like me, so I fetched a large looking-glass to make sure, and then to my great joy I found it was Me. We shook hands, and were just beginning to talk, when Myself came up and joined us, and we had quite a pleasant conversation. I said, 'Do you remember when we all met at Sandown?' And Myself said, 'It was very jolly there; there was a child called Magdalen,' and Me said, 'I used to like her a little. Not much, you know—

[1] *The Hunting of the Snark* was not published until the spring of 1876.

> only a little.' Then it was time for us to go to the train—and who do you think came to the station to see us off? You would never guess, so I must tell you. They were two very dear friends of mine, who happen to be here just now, and beg to be allowed to sign this letter as your affectionate friends, Lewis Carroll and C. L. Dodgson."

The late Mr. Dodgson was a clergyman in deacon's orders. The reason why he was never ordained a priest was said to be a slight hesitancy of speech which prevented his speaking in public. This, however, he in a measure overcame, and he not infrequently read the lessons and prayers at the college services in Oxford Cathedral. He had even occasionally been known to preach at the special services for college servants, but it was very rarely that he mounted the pulpit. He was a creature of habit, and in term time was never absent from his own particular seat in St. Mary's for the University sermon, always staying to matins afterwards. During the last five-and-twenty years he had hardly ever been missed from his accustomed place. Until latterly he was accustomed to take long walks alone.

Tribute by Canon Sanday at Oxford Cathedral.[1]

At the Cathedral on Sunday morning Canon Sanday, Margaret Professor of Divinity, made a feeling reference in his sermon to Mr. Dodgson's death. He said it fell to him to speak of the unexpected and saddening news which reached them only the previous day, and which would cast its shadow over the approaching term. The term would come, but they would no more have among them the living presence of one who for all the retirement in which he lived had for many years made Christ Church his home, and could not help filling no small place in the picture which the world without formed to itself of Christ Church. The outer man was familiar to most of those who frequented their streets and rambled through the lanes of the country round. The inner man was revealed in many a charming book, and in fact was so transparent that even the ordinary meetings of daily life, with an occasional walk and talk, were enough to reveal its main outlines. All that made the individual, the infinite play of fancy and the subtle undercurrents of serious and chastened thought must needs be lost to them; they went with him whose they were to inhabit another sphere than ours. The world would think of "Lewis Carroll" as one who opened out a new vein of literature, a new and a delightful vein

---

[1] The following portion of this article is a reprint of a piece which appeared in *The Times* on 17 Jan. 1898, under the title "Lewis Carroll."

which added at once mirth and refinement to life. Was not this much to have done? Did it not bespeak an original and creative gift of a high order and of great value in the eyes of those who had the best interests of the nation at heart? His books were wholly without spot or stain in the midst of so much that was the reverse of stainless, and were all seasoned with such a wholesome salt of humour as won for them a wide popularity. Might they not say that from their courts at Christ Church there had flowed into the literature of their own time a rill bright and sparkling, healthgiving, and purifying wherever its waters extended? Perhaps there were few happier spots in the English literature of the century than its books for children, and how large a share of this was due to him who was gone! For this all England was grateful, nay, more, all readers of the English tongue. But they in this place knew how fully the man bore out the promise of his books. They knew what an interest he took in an art which offered great scope for genius and great capacity for good, though beset with not a few temptations. They knew what a high and scrupulous standard he brought to this, and how excellent was his influence in regard to it. They knew his fondness for children, and what trouble he took to make them happier and better. But, most of all, they knew what was the fount and spring from which all these varied activities took their direction. They knew how behind them all there lay a deep background of religion—a religion severely quiet and retiring, like his character—a religion almost of the closet after the pattern of the Gospel. They knew that his public utterances were rare but striking in their rarity. They knew how he chose out by preference the smallest and least frequented of their services, and they knew how when there the bowed head and close concentrated attention spoke of a soul seeking communion with God. From these things they must not further lift the veil.

The Funeral[1]

In conformity with his written instructions, the funeral of the Rev. Charles Lutwidge Dodgson (Lewis Carroll) was carried out at Guildford on Wednesday in the quietest manner. The coffin, which was covered with flowers, was conveyed on a hand-bier to the parish church of St. Mary, where the first portion of the Burial Service was conducted by the Dean of Christ Church (the Very Rev. Dr. Paget), assisted by Canon Grant (rector), the Rev. A. Kingston and the Rev. E. C. Kirwan (curates of the parish). The mourners were the Rev. S. H. Dodgson and Mr. W. L. Dodgson (brothers), Miss Elizabeth Dodgson, Miss Caroline Dodgson,

---

[1] The following portion of this article is a reprint of a piece which appeared in *The Times* on 20 Jan. 1898.

and Miss Margaret Dodgson (sisters), Mr. C. H. W. Dodgson (nephew), Mr. J. H. D. Dodgson (cousin), and Mrs. Wilcox, of Winchester (cousin). Among others present at the service were the Rev. T. J. Prout, the Rev. T. B. Strong, Mr. A. G. V. Harcourt, and Mr. A. Hassall (all of Christ Church), Mr. G. L. Craik (representing Messrs. Macmillan), Dr. Gabb, the Rev. H. R. Ware, Dr. Clark, and Mr. G. C. Williamson. The hymns sung at the service by the choir were "The Saints of God" and "My God, my Father, while I stray." On leaving the church the coffin was placed on the hand-bier and conveyed to the cemetery, followed by the mourners on foot. The committal sentences were read at the graveside by the Dean of Christ Church, and the choir sang "Peace, perfect peace." Among the wreaths were several from relatives, and others from the Dean of Christ Church and Mrs. Paget, the members of the Common Room, Christ Church, Mr. J. H. Arkwright, Mr. and Mrs. R. E. Baynes, Mrs. Reginald Hargreaves, and Miss Weld.[1] The coffin bore the inscription, "Charles Lutwidge Dodgson, born January 27, 1832, died January 14, 1898."

Oxford Journal
January, 1898

All will lament the death of the Rev. C. L. Dodgson, Student of Christ Church, who died at Guildford at the age of sixty-six. Mr. Dodgson was "Lewis Carroll," but as Lewis Carroll he has been adequately and appreciatively noticed in many newspapers. As Mr. Dodgson he was a member of Christ Church, formerly Mathematical Tutor, latterly a Student (that is, Fellow), holding his studentship for life like the older Fellows of all colleges, and residing regularly in Christ Church during term. He was old when he died, and as years advanced he had tended to retire into himself, but he was always a man of the utmost courtesy. His graceful and often brilliant conversation and his pleasant humour charmed all who met him in Christ Church Hall or elsewhere. His jokes and stories, often based on turns of language and logical or mathematical fancies, bridged over a gap which seems to have puzzled some of his biographers, and explained how the creator of *Alice* could also have been a mathematician who wrote serious treatises on a learned subject. Of his mathematical powers I cannot speak. I have been told that he was an excellent teacher but less successful as a writer on higher

---

[1] For a more complete, footnoted list of mourners, see below pp. 157–8.

mathematics. But he always kept clear the dividing line between Mr. Dodgson and Lewis Carroll; there was no surer way for a stranger to his disfavour than to ask for his autograph, or to allude to the books which he wrote as "Lewis Carroll." Though ordained, he took little active part of late years in spiritual work, but he was always a man of profound religious feeling, and he especially objected to anything which even savoured of a joke on a sacred subject.[1] Recently he had preached two striking sermons to undergraduates in St. Mary's,[2] and we had hoped to hear him oftener. But it was not to be. Pneumonia following on influenza suddenly carried him away from us, full of years and work and honour. He will be buried at Guildford on Wednesday.

The Oxford Magazine
January 26, 1898

Illustrious and well known as certain Christ Church names are, yet, as the Dean of Christ Church well said in his eloquent sermon, there were probably hardly two names better or more widely known, in the realms of letters at least, than the two which the House must now transfer from the record of her present to that of her past ornaments. *Liddell and Scott* and *Alice in Wonderland*—Dr. Liddell and Mr. Dodgson: it seemed strange that they should have been brought into such close conjunction in life, and in death they were not divided. The world perhaps hardly knew that the Dean was a great artist, or that "Lewis Carroll" was a great mathematician. The fact is they were both men of remarkably original and individual gifts, and even genius; such as are rarely lost by, because rarely given to, one College at the same time.

To "Lewis Carroll" the papers generally have done very fair justice, except that they have not paid sufficient attention to his academic skits, such as those on the *Evaluation of* π, or the meat-safe, some of the wittiest and funniest things of the kind ever written. We give elsewhere an extended notice by one who knew and loved him well.

---

[1] Dodgson explained this attitude in his article "The Stage and the Spirit of Reverence," *The Theatre* (June 1888), 285–89.

[2] On 6 Dec. 1896, Dodgson preached at the evening service at St. Mary's, the University Church, taking as his text Mark 9:24. His two sermons preached at St. Mary's at the special evening services for undergraduates were on 7 March 1897 when he preached on Job 28:28 "for about three quarters of an hour" and on 24 Oct. 1897, when he preached (on Psalm 103:11–12) a sermon on reverence.

To the Dean they have done perhaps less than full justice. *Liddell and Scott* is a pedestal upon which he will stand for many a year, and will tower higher as he recedes into the past. But he was not only an artist, a scholar, a man of letters, but had also great practical gifts for affairs. It was said, and with justice, that he was the best Chairman in Oxford. We hope next week to publish a full account of him, also by an old friend well qualified to speak.[1]

The Oxford Magazine[2]
January 26, 1898

It is not easy to estimate the loss which Oxford has sustained in the unexpected death of Mr. Dodgson. It might almost be said of him, as of another famous Christ Church Student of somewhat kindred genius, that he was *paucis notus, paucioribus ignotus*.[3] As "Lewis Carroll" he had a world-wide reputation; but at Oxford he took no part in University business, and was seldom seen in general society. His habits were in some respects those of a recluse, and he had not many intimate friends. Yet those who were privileged to share his friendship know that their lives are now far poorer than before, and that there has passed away one of the most delightful of companions, whose keen intellect and playful fancy, united to a guilelessness and purity almost childlike in their simplicity, gave a rare and unique charm to a friendship now to be cherished only in memory.

The Rev. Charles Lutwidge Dodgson was the son of Archdeacon Dodgson, a former Westminster Student of Christ Church, a brilliant and versatile mathematician and scholar, whose Latin epilogues and epigrams are still worth reading. He was appointed by Dr. Pusey to a Studentship on the old Foundation, which he relinquished in due course. But having been created a Senior Student under the Ordinance of 1858, he was able to retain his position as an "Alumnus."[4] After distinguishing himself in the Final Schools he was appointed Mathematical Lecturer at Christ

---

[1] This obituary, by H. L. Thompson (see note below) did appear on 3 Feb. 1898.

[2] This article, titled "The Late Rev. C. L. Dodgson," was signed "H. L. T." and written by Henry Lewis Thompson (1840–1905), Student and Censor at Christ Church and friend of Dodgson. Thompson also wrote a biography of Dean Liddell (1899) and a history of Christ Church (1900).

[3] "Known by few, but by fewer not known."

[4] The Ordinance of 1858 created two classes of Studentships, Senior Studentships, tenable for life; and Junior Studentships, tenable for five or seven years.

Church in 1855, and held that post till 1881. From that time he took no part in the tutorial work of the College, but he continued to reside within its walls till his death, living in rooms on the first floor at the north-west corner of the Great Quadrangle, and occupying himself incessantly with literary labour of the most varied kind.

Of his work as a mathematician a layman must not venture to hazard an opinion. He was a strenuous student, and an independent thinker: and in *Euclid and his Modern Rivals* he is believed to have achieved his object, in ousting from public school teaching one at least of unsound modern manuals. Of late, he was profoundly interested in logical studies; and if his *Symbolic Logic* had been completed, it would certainly have merited the praise of ingenious and original treatment of a well-worn subject, encumbered perhaps with trivialities which sometimes, as they were propounded to bewildered friends during a country walk, recalled the useless puzzles of ancient sophistry.

His special gifts of fancy and humour, delicate, quaint, and quite inimitable, gave a character of its own to all that he touched. He used from time to time to contribute to the passing controversies of the University or Christ Church some infinitely humorous leaflet, which softened all bitterness and dissolved animosities in endless laughter. *The New Method of Evaluation as applied to π*, is now perhaps almost forgotten. It was a paper circulated in 1865, at the conclusion of the long controversy about the endowment of the Regius Professorship of Greek, and it described, under mathematical formulae, the various methods attempted for attaining the desired result. In 1868, when the Professor of Physics met an offer of the Clarendon Trustees by a detailed enumeration of the requirements in his own department of Natural Science, Mr. Dodgson, in a "Letter to Sandford,"[1] using almost the exact phraseology of the Professor, described in a very amusing way the imaginary needs of the Mathematical School. He asked among other things for "a room for reducing Fractions to their Lowest Terms. This should be provided with a cellar for keeping the Lowest Terms when found, which might also be available to the general body of Undergraduates for the purpose of 'Keeping Terms,'" and "a narrow strip of ground, railed off and carefully levelled, for investigating the properties of Asymptotes, and testing practically whether Parallel Lines meet or not: for this purpose it should reach, to use the expressive language of Euclid, 'ever so far.'"

It is impossible to explain fully the secret of his charm as a writer for

---

[1] *The Offer of the Clarendon Trustees* (1868) was a parody of a letter from Robert Bellamy Clifton to Charles W. Sandford, the Senior Censor of Christ Church.

children. *Alice in Wonderland*, and *Through the Looking Glass* almost defy analysis or criticism. But they were the outcome of a very beautiful life; and the insight into the mind of childhood no doubt came in a great measure from the singular innocence and refinement of his own mind, with its strong background of reverence and awe. He was in truth a deeply religious man, almost a puritan in strictness; always on the guard against the intrusion into conversation of any expression which savoured in the least degree of light estimation of Divine truth. Some will remember his sermon at St. Mary's last Lent Term; the erect gray-haired figure, with the rapt look of earnest thought; the slow, almost hesitating speech; the clear and faultless language; the intense solemnity and earnestness which compelled his audience to listen for nearly an hour, as he spoke to them on the duty of reverence, and warned them of the sin of talking carelessly of holy things.

He was then revealing the secret of his life; and those who were his friends are thankful for his friendship, and desire, now that he has passed away, To shed streams of tears in memory of his kindness.[1]

The Oxford Review
January 25, 1898

Whatever the exact number of copies of *Alice in Wonderland* in circulation may be it is impossible to say without more accurate knowledge than an outsider can possess, but there is no doubt it has been one of the most widely-read books of the century. There has been much writing concerning the late Mr. Dodgson during the last few days, but the following facts may be fresh to many of our readers. Mr. Dodgson was a member of a large family, and at one time it was their custom to have a family paper, got up among themselves for home circulation, to which each member contributed something—one a sketch, another some rhymes, and the third a story. "Lewis Carroll," in addition to acting as editor, took his share in writing, and it is probable that some at least of the adventures of *Alice in Wonderland* originally took a tangible form in this way[2]. Mr. W. L. Dodgson[3] has still in his possession the copies of

---

[1] This phrase was printed in Greek. It is from Meleager, *Greek Anthology, VII.476.4*

[2] Dodgson and his family created eight family magazines in this fashion. In the four that survive there is little material which appears later in the *Alice* books. The most significant "Alice" item from these magazines is the "Stanza of Anglo-Saxon Poetry" from *Mischmasch*, which became the first stanza of "Jabberwocky."

[3] Wilfred Longley Dodgson (1833–1914). Charles Dodgson's younger brother.

this little journal, which is full of quaint things. It is written on sheets of notepaper and stitched together. It was compiled as follows: The editor, having obtained the blank sheets, wrote his contribution, and apportioned the amount of space to be filled by each member of the family. It was then put in circulation, and after each individual had filled up his space the completed journal was duly "published" and sent round once more for perusal. Some of the sketches, representing animals of the same family as the "Snark" and the "Boojum," were truly appalling.

A little knowledge is said to be a dangerous thing, both to its possessor and the unoffending public. A young man who was coming up the other day was a striking example. A friend who was with him remarked, "Dodgson is dead." "Oh, yes," said the other. "Lewis Carroll, you know," said the first. "What did he write?" "*Alice in Wonderland*, of course." "Ah," said the young man with a smile of lofty superiority, "I read it in the nursery," with which he dismissed the subject from his mind. If the gentleman had read it in the nursery it may account for his want of appreciation, but his surprise at the interest other people took in the book was very amusing. That the late Mr. Dodgson took a tremendous interest in all young people and their doings is well known. He also interested himself on behalf of young actors and actresses starting their profession, and many is the kind letter he has written them. To mention even a small portion of the things that he wrote and said to amuse the children of his acquaintance would fill columns. Some men have not the good fortune to be appreciated by their friends, but all the members of the Dodgson family, young and old, looked eagerly forward to a visit from their "Uncle Charles" as a red-letter day in their calendar.

The Oxford Times[1]
January 22, 1898

The Rev. C. L. Dodgson, who under the pseudonym of "Lewis Carroll" produced *Alice in Wonderland* and *Through the Looking-glass*, died on Friday at Guildford, in his 66th year. The Rev. C. L. Dodgson was a distinguished mathematician, senior student of Christ Church, Oxford, in deacon's orders in the Church of England, and for a

---

[1] Portions of this obituary also appeared in *The Oxford Chronicle* (22 Jan. 1898). See above pp. 109–15.

quarter of a century mathematical lecturer at Christ Church. In examining a list of his published works it is curious to find *Alice's Adventures in Wonderland* sandwiched between *The Formulæ of Plane Trigonometry* and *An Elementary Treatise on Determinants*; *Through the Looking-Glass* followed by *Euclid, Book V. proved Algebraically*; and *The Hunting of the Snark* by *A New Theory of Parallels* and *Symbolic Logic*, with last, and possibly least, *Silvie and Bruno*. It is related that when the Queen read *Alice's Adventures* she was so charmed that she commanded that all his writings should be procured for her perusal, and was disappointed to find that the majority of them were but text-books of mathematical theme.

At the Cathedral on Sunday morning, Canon Sanday, Margaret Professor of Divinity, was the preacher and made a feeling reference in his sermon to the death of "Lewis Carroll." He said that for all the retirement in which he lived Mr. Dodgson had for many years made Christ Church his home, and could not help filling no small place in the picture which the world without formed itself of Christ Church. The outer man was familiar to most of those who frequented their streets and rambled through the lanes of the country round. The inner man was revealed in many a charming book and in fact was so transparent that even the ordinary meeting of daily life, with an occasional walk and talk, were enough to reveal its main outlines. All that made the individual, the infinite play of fancy and the subtle undercurrents of serious and chastened thought must needs be lost to them. They went with him whose they were to inhabit another sphere than ours. The world would think of "Lewis Carroll" as one who opened out a new vein in literature, a new and delightful vein which added at once mirth and refinement to life. Was not this much to have done? Did it not bespeak an original and creative gift of a high order and of great value in the eyes of those who had the best interest of the nation at heart? His books were wholly without spot or stain in the midst of so much that was the reverse of stainless, and were all seasoned with such a wholesome salt of humour as won for them a wide popularity. Might they not say that from their courts at Christ Church there had flowed into the literature of their own time a rill bright and sparkling, healthgiving, and purifying wherever its waters extended? Perhaps there were few happier spots in the English literature of the century than its books for children, and how large a share of this was due to him who was gone! For this all England was grateful, nay, more, all readers of the English tongue. But they in this place knew how fully the man bore out the promise of his books. They knew what an interest he took in an art which offered great scope for genius and great capacity for good, though beset with not a few temptations. They knew what a high and scrupulous standard he brought to this, and how excellent was his influence in regard to it. They knew his fondness for children and what trouble he took to make them happier and better. But, most of all, they knew what was the fount and spring from which all these varied activites took their direction. They knew how behind them all there lay a deep background of religion—a religion severely quiet and retiring, like his character—a religion almost of the closet after the pattern of the Gospel. They knew that his public utterances were rare but striking in their rarity. They knew how he chose out by preference the smallest and least frequented of their services, and they knew how when there the bowed head and close concentrated attention spoke of a soul seeking communion with its God. From these things they must not further lift the veil.

In conformity with his written instructions, the funeral of the Rev. Charles Lutwidge Dodgson was carried out at Guildford in the quietest manner.[1] The coffin, which was covered with flowers, was conveyed on a hand-bier to the parish church of St. Mary, where the first portion of the Burial Service was conducted by the Dean of Christ Church (Dr. Paget), assisted by Canon Grant (rector), the Rev. A. Kingston and the Rev. E. C. Kirwan (curates of the parish). The mourners were the Rev. S. H. Dodgson and Mr. W. L. Dodgson (brothers), Miss Elizabeth Dodgson, Miss Caroline Dodgson, and Miss Margaret Dodgson (sisters), Mr. C. H. W. Dodgson (nephew), Mr. J. H. D. Dodgson (cousin), and Mrs. Wilcox, of Winchester (cousin). Among others present at the service were the Rev. T. J. Prout, the Rev. T. B. Strong, Mr. A. G. V. Harcourt, and Mr. A. Hassall (all of Christ Church), Mr. G. L. Craik (representing Messrs. Macmillan), Dr. Gabb, the Rev. H. R. Ware, Dr. Clark, and Mr. G. C. Williamson. The hymns sung at the service by the choir were "The Saints of God" and "My God, my Father, while I stray." On leaving the church the coffin was placed on the hand-bier and conveyed to the cemetery, followed by the mourners on foot. The committal sentences were read at the graveside by the Dean of Christ Church, and the choir sang "Peace, perfect peace." Among the wreaths were several from relatives, and others from the Dean of Christ Church and Mrs. Paget, the members of the Common Room, Christ Church, Mr. J. H. Arkwright, Mr. and Mrs. R. E. Baynes, Mrs. Reginald Hargreaves, and Miss Weld. The coffin bore the inscription, "Charles Lutwidge Dodgson, born January 27, 1832, died January 14, 1898."

---

[1] For a footnoted list of funeral participants, see pp. 157–59.

Oxford University Gazette January 21, 1898

Died, on Friday, January 14, at Guildford, the Rev. Charles Lutwidge Dodgson, M.A., Student and sometime Mathematical Lecturer of Christ Church. Aged 65.

Pall Mall Gazette (London)[1] January 15, 1898

The loss of the Rev. Charles Lutwidge Dodgson, M.A., will sadden dozens of friends; that Lewis Carroll will be mourned by thousands and thousands who never saw him, and very possibly did not even know that his name was Dodgson. The remark that there will be sorrow in many a schoolroom and many a nursery was inevitable;[2] but it would have been truer if "and wherever there is an educated man" had been added. The late Master of Balliol used to say that nobody could call himself educated who had not read his Boswell,[3] and the same may be said of any who have not read their *Alice*. If your parents are wise, you first read of her when you are five, six, or seven, and if you are wise yourself, you are still reading of her at fifty, sixty, or seventy, for the fortieth or fiftieth time. It is pardonable to know little of Mr. Dodgson. A dichotomy into two men is the phrase in speaking of him, though it is possible to discover at least three if you care to, for the Rev. Charles Dodgson should not be lost sight of. No man who had depicted the child Alice could have been other than a good man, but many will be interested to hear that he was a clergyman of unimpeachable orthodoxy, a man of deep piety, and one who devoted much of the profits of his works to the charitable causes in which he took such intimate interest. Yet the striking contrast is between Mr. Dodgson, the mathematician of "the House," of the dons donnish, an ice-bath to the undergraduate of the slacker sort, and the Lewis Carroll of the books and of private life—for it is better to call that charming, witty, lighter side of him by the well-loved name. As a rule, the two were kept distinct, and a confusion of them was

---

[1] Dodgson contributed several items to *Pall Mall Gazette* between 1866 and 1877.

[2] This remark appeared in both *The Daily News* (15 Jan. 1898) and *The Times* (15 Jan, 1898), and was repeated in *Alliance News* (6 May, 1898). See above pp. 33, 162, 18.

[3] James Boswell (1740–1795). Author of a biography of Samuel Johnson.

not always fortunate. Her Majesty for one may have thought so, if the old story be true that she was so pleased with *Alice in Wonderland* that she commanded a copy of his next work, and was astonished to get *An Elementary Treatise on Determinants*. The *Game of Logic*, again, it must be confessed, is too serious for fun and too light for business; it bores a child and is only a not very living sort of formal logic at best. Let it be granted, however, that Mr. Dodgson's delightfully humorous essays on University politics, and his parody of Euclid, with its definition of "a plain, superficial speaker" as "one who, any two points being given, will lie evenly between those two points," and so forth, make rather against us. It was in Oxford also that many of Lewis Carroll's best things in private conversation were said, and that he derived the inspiration for Alice from one of the daughters of the Dean of Christ Church.

Of the former many delightful specimens are given to-day, the prize one, perhaps, the *Telegraph*'s, his reply to the tandem-driving undergraduate who offered him a lift, having on the day before spilled the Khedive, who was visiting Oxford: "Wilt thou slay me as thou didst the Egyptian yesterday?" Charming pictures also are given of the Paradise his rooms presented to a child. But after all Alice and the Snark are the thing. We are always quoting them, all of us. "For the Snark was a Boojum, you see," and "Jam yesterday, and jam to-morrow, but never jam to-day," come in at every other turn. Yesterday a correspondent of the *Times* cited the anonymous document portion of the trial scene in "Wonderland" as pat for the Dreyfus case; we wonder if the *Times* remembers that that same passage was applied some years ago to the readiness with which certain people caught at a letter attributed to Mr. Parnell.[1] No one would thank us for attempting to analyze the beauties of Lewis Carroll's classics. It is well, however, to insist that one supreme excellence lay in the fact that they never moralized—which is exactly why *Sylvie and Bruno*, which did, fell as flat as the gardeners before the Queen of Hearts. Again and again a sort of idea floats before one that there is something at the back of all this if one could only get it, but one never does. There were people, for instance, who argued with most ingenious corroborative detail that Popularity was the real name of the Snark—which, by the way, we believe, was written up to the final "For the Snark was a Boojum, you see," a line which flashed into Lewis Carroll's head one day on a country walk. A whole metaphysical problem seems to underlie the question whether it is really the Red King who is dreaming or ourselves; and where

[1] Charles Stewart Parnell (1846–91). Irish political leader.

is there a more significant figure than his greatest creation, the poor, pathetic White Knight, though the significance be of the most undefined? The more we read of Lewis Carroll, the more we find; and yet the youngest child is more than contented with the pure nonsense at the first reading. The man who created the Mock Turtle, the Walrus and the Carpenter, and the Mad Tea-party alone, would have made his own fame; the portrayal of the child Alice showed the thorough comprehension of genius. He has had imitators in plenty, some few good, but none has rivalled or will rival him.

The Pall Mall Gazette (London)[1] January 17, 1898

The late "Lewis Carroll" (the Rev. Charles Lutwidge Dodgson), who has just died at the age of sixty-six, did not really begin his literary career in 1860, by the publication *of A Syllabus of Plain Algebraical Geometry*, but a few years earlier, with much lighter productions, in the pages of the *Train Magazine*. This magazine was started and edited by the late Edmund Yates on something like co-operative lines. The various contributors furnished stories, essays, poems, and jokes, mostly signed with their names, making their chance of payment depend upon the financial success of the magazine. *The Train* had only one fault, but that was a great and serious one. It was ten or fifteen years before its time. Its price was a shilling; it was a monthly magazine, its publishers were an old-fashioned firm in Paternoster-row, now obliterated, and its co-operative owners had little or no capital to "push" and develop the publication. Edmund Yates was then a hard-working young beginner, engaged in the day-time at the General Post Office, and at night on the *Daily News*, the *Illustrated Times*, and other journals, occasionally writing a farce in collaboration with his fellow-clerk, Harrington ("My Friend from Leatherhead," for example), and getting it performed at a London theatre. The co-operative contributors to the *Train* which was started in January, 1856, were writers as well known and popular as George Augustus Sala, Robert Brough, Frank E. Smedley, Edward Draper, and Godfrey Turner (born humourists), W. P. Hale (the son of Archdea-

[1] This article, titled "Lewis Carroll and a Forgotten Magazine," was signed by "John Hollingshead," playwright and author of *The Birthplace of Podgers*, which Dodgson saw on 21 Aug. 1891.

con Hale), Palgrave Simpson, William Brough, John V. Bridgman, Sutherland Evans, J. Deflett Francis, J. H. Friswell, Albany Fonblanque, William M. Thomas, Thomas Archer, Andrew Halliday, myself (then a young beginner),[1] with artists like C. H. Bennett, W. McConnell, and others. The editor wrote much and well, and kept his contributors well together. Mr. "Lewis Carroll" turned up soon after the appearance of the first number, and his contribution appeared in March, 1856, some sentimental verses called "Solitude." His second contribution was a light article in prose, which was called "Novelty and Romancement," and appeared in the October number of 1856. In the next number—that of November, 1856—a long poem of eighty-four three-line verses was published, called "The Three Voices." It was written as a refined parody of Tennyson's "Two Voices," and gave unmistakable indications of power and humour. Mr. "Lewis Carroll" at that time was about twenty-four years of age, and had not yet been ordained a deacon. In May, 1857, he was again represented in the *Train* by a serious poem called "The Sailor's Wife," illustrated by C. H. Bennett, and in the December of the same year appeared a clever parody of Longfellow's "Hiawatha." "Lewis Carroll" was known to be an expert amateur photographer, and this poem was called "Hiawatha's Photographing." An apologetic note precedes the poem, stating how easy the author found it to imitate the professor's metre.[2] The poem describes all the stages of the ordinary photographic process, forty years ago, evidently drawn with knowledge. The humour is shown in the description of the family "taken." The poem is too long to quote, running to nearly 180 lines, but a picture of one "sitter" may be given as a specimen:—

---

[1] Primarily minor writers of the 1850s, including: George Augustus Sala (b. 1827), contributor to *Dickens' Household Words* and editor of *Temple Bar* (to which Dodgson contributed a poem in 1861); Francis Edward Smedley (1818–64), novelist (author of *Frank Farleigh*), poet and distant cousin of Dodgson who helped him to get some early poems published in Yates' *Comic Times*; William Palmer Hale, matriculated Christ Church 1842, practiced law in London from 1852; J. Palgrave Simpson, author of *Second Love, and Other Tales* (1846), *Pictures of Revolutionary Paris* (1848) and others; Albany Fonblanque (b. 1800), proprietor and editor of the *London Examiner*, author of *England Under Seven Administrations* (1837); William Moy Thomas (b. 1828), literary, political, and dramatic critic, contributor to numerous papers (including *Household Words* and *The Pall Mall Gazette*) and columnist for *The Daily News*.

[2] Dodgson wrote "Any fairly practised writer, with the slightest ear for rhythm, could compose, for hours together, in the easy running metre of 'The Song of Hiawatha.'" Longfellow was a professor at Bowdoin and at Harvard.

First, the governor—the father—
He suggested velvet curtains
Looped about a massy pillar,
And the corner of a table—
Of a rosewood dining-table.
He would hold a scroll of something—
Hold it firmly in his left hand;
He would keep his right hand buried
(Like Napoleon) in his waistcoat;
He would gaze upon the distance—
(Like a poet seeing visions,
Like a man that plots a poem,
In a dressing-gown of damask,
At 12.30 in the morning,
Ere the servants bring in luncheon)
With a look of pensive meaning,
As of ducks that die in tempests.

The Train died itself in the following June, not in a tempest, but a dead calm, having completed its fifth volume and two and a half years' existence. "Lewis Carroll" may or may not have reprinted these early trifles in "Phantasmagoria," published in 1869, but it will certainly help to complete the literary history of a great writer to call attention to them.[1]

The Philadelphia Item[2] January 15, 1898

The death of Rev. Charles Lutwidge Dodgson, author of that famous child's book, *Alice in Wonderland* is commented upon at length in all to-day's journals.

The Times declares that Mr. Dodgson was to England what Hans Christian Anderson was to the world in general.

---

[1] Six of Dodgson's eight contributions to *The Train* were reprinted in *Phantasmagoria*, and some also appeared in *Rhyme? and Reason?* (1883) and *The Three Sunsets* (1898).

[2] Excerpts from this article appeared in the *Pittsburgh Dispatch* (15 Jan. 1898) under the title "Children's Author Dead. Rev. Charles L. Dodgson, Who Wrote *Alice in Wonderland*—Could Not Comprehend His Own Fame," with slightly different opening three sentences.

The Rev. Charles Lutwidge Dodgson, M.A. ("Lewis Carroll"), was born in 1830 and was graduated from Christ Church College, Oxford, in 1854. From 1855 to 1881 he was mathematical lecturer at Christ Church. The man who wrote *The Hunting of the Snark* also wrote *A Syllabus of Plane Algebraical Geometry*, and contributed many works to the literature of mathematics.

He never could realize that fame had come to him, not as an advanced mathematician, but as the author of the most fascinating nonsense that ever was written:[1] *Alice in Wonderland, Alice Through the Looking-Glass, The Hunting of the Snark, Agony in Eight Fits, Sylvie and Bruno*, and the rest of it.

When she first made her bow and Alice's wonderful adventures were the delight of children of every growth, Queen Victoria sent a message to "Lewis Carroll" asking him to send her his next book. Like every one else, Her Majesty was anxious to know more of the delightful child whose prototype was the daughter of the Dean of Christ Church, now a woman most happily married. The Queen was much astonished to receive a copy of *An Elementary Treatise on Determinants*, by C. L. Dodgson. In those days "Lewis Carroll" managed to preserve his incognito, and the Queen, like the rest of the world, believed him to be a mere humorist.

Mr. Dodgson was tall and spare, with a clean shaven face, half humorous, half melancholy. His manner was shy, yet precise. He was a clergyman in deacon's orders; he was never ordained a priest owing, it was said, to a slight hesitancy in his speech which prevented his speaking in public.

He dearly loved children, just as children loved him, whether they knew him or not. He delighted in the dramatic version of *Alice* chiefly because it was played entirely by children.[2] "Lewis Carroll" was in the habit of sending gifts to all the diminutive company.

Mr. Dodgson had delightful rooms in "Tom's Tower,"[3] Christ Church College. They were crowded with interesting momentoes; his bookshelves teemed with rare first editions. He was an ardent amateur photographer.

---

[1] Dodgson certainly realized that his nonsense works were the source of his fame. Whether he liked that fact is a matter for dispute.

[2] Although the cast of Savile Clarke's production of *Alice* was made up primarily of children, there were some adults in the company as well.

[3] His rooms (from 1868–98) were in the northwest corner of Tom Quad, near Tom Tower (see above p. 29).

The Providence Journal (Providence, Rhode Island) January 17, 1898

The death of the author of *Alice in Wonderland* removes from English writers one who sought distinction in the solid ways of science and found it where he never looked for it. He will be long remembered for his children's books. It may be that *Alice in Wonderland* and *Through a Looking Glass* will be classics, to be forgotten only when such plays of fancy as "The Spider and the Fly,"[1] which was written, it is supposed, as an allegory, have been put one side by a generation not requiring such media for its instruction. Rev. Mr. Dodgson had no more serious idea in presenting his children's books to the world than that they should amuse young minds in idle hours. He did not produce them as the principal work of his life. Yet it is perhaps from that very fact that they derive a perfection of artistic severance from everyday existence. It may account for their magical charm over adult minds. Certainly the race of the Anglo-Saxons, though prone to be serious in all ways, is possessed of a strong love of the nobly fanciful and the detached and dreamy.

Yet in the truest sense Lewis Carroll was not a two-sided man. Like Pascal[2] he found his chosen tasks too confining and narrowing for his mental nature. Unlike him, in the course of the hard work of a lifetime he demanded a separate outlet for the portion of his inner personality which had to have expression. Late in his career he united mathematics and humor in a book which his real friends read with some pleasure, though no fascination.[3] There one sees the creator of the all but immortal Alice as a humorist. The lovers of genuine humor, however, can detect this quality in his two most famous books, although in places it is not spontaneous. But as a whole the stories are richly human, heart-warming and fondly delusive with their grave but keen wit and their clear evidences of the author's perception of all that underlies the hard crust of even adult thinking in this workaday world. This is the real Lewis Carroll, in the crucible of whose mind the exact sciences and the love of the fanciful could not be wholly fused. The natural cast of his nature was too inflexible to let him use his noble humor to good effect upon Euclid and Hegel[4].

---

[1] Dodgson's poem "Will You Walk a Little Faster," from *Alice*, is a parody of the poem "The Spider and the Fly" by Mary Howitt (see Gardner, *The Annotated Alice*).

[2] Blaise Pascal (1623–62), French philosopher, mathematician, and scientist.

[3] This may refer to *Symbolic Logic—Part I Elementary* (1896).

[4] Georg Wilhelm Friedrich Hegel (1770–1831), German idealist philosopher.

In general talk he is always classed as a writer of fairy stories. But he was distinctively a creator of fancies, not of fairies. He never approached the fantastic. He stood so firmly upon the field of pure fancy which he created, or discovered, for himself, that the reader always remains fixed upon an actual conception of the convenances of natural life. Here he was as artistic as Andersen in his dual realm of fairyland and fantasy. George Sand[1] alone has imparted to fairy stories the same element of subjectivity for each reader. Even if the book which is said to be a satire on public men and affairs in England is such, the effectual detachment of the story from the purpose is sufficient to rob the most curious of any real desire to trace out the connection. The incapacity of the author for the fantastic must be evident in this instance, for *The Hunting of the Snark* is bold enough in its conception to uphold considerable supernatural machinery. But Lewis Carroll was a portrayer of fancy alone in the world of humor. His characters, if such they be, have no interest or concern for us derived from anything but what is in our own minds.

Punch (London)[2]
January 29, 1898

Lover of children! Fellow-heir with those
Of whom the imperishable kingdom is!
Beyond all dreaming now your spirit knows
The unimagined mysteries.

Darkly as in a glass our faces look
To read ourselves, if so we may, aright;
You, like the maiden in your faerie book—
You step beyond and see the light!

The heart you wore beneath your pedant's cloak
Only to children's hearts you gave away;
Yet unaware in half the world you woke
The slumbering charm of childhood's day

[1] Pseudonym of Aurore Dudevant (1804–76), French novelist.

[2] This unsigned poem appeared under the heading "'Lewis Carroll.' Born 1832. Died January 14, 1898."

We older children, too, our loss lament,
    We of the "Table Round,"[1] remembering well
How he, our comrade,[2] with his pencil lent
    Your fancy's speech a firmer spell.

Master of rare woodcraft, by sympathy's
    Sure touch he caught your visionary gleams,
And made your fame, the dreamer's, one with his,
    The wise interpreter of dreams.

Farewell! But near our hearts we have you yet,
    Holding our heritage with loving hand,
Who may not follow where your feet are set
    Upon the ways of Wonderland.

St. James's Gazette (London)[3]
January 15, 1898

With deep and peculiar regret we have to record the death, which took place yesterday at The Chestnuts, Guildford, the residence of his sisters, of the Rev. Charles Lutwidge Dodgson, the famous—it would, perhaps, not be extravagant to say the immortal author of *Alice in Wonderland.* The name of "Lewis Carroll" was familiar to hundreds of thousands who had never heard of the Rev. Mr. Dodgson, the distinguished mathematician, and the regret which will be felt at his death will be shared by the whole English-speaking world, the more especially since,

---

[1] The writers and artists of *Punch* met on Wednesday evenings for an editorial dinner at a round table to plan the magazine's content.

[2] This refers to John Tenniel, illustrator of the *Alice* books and still, in 1898, cartoonist for *Punch*. Tenniel drew his illustrations on wood for engraving, hence the reference to the "Master of rare woodcraft" below.

[3] Portions of this obituary appeared in *The Daily Telegraph* (15 Jan. 1898). See above pp. 37–40 for additional footnotes. Dodgson was a frequent contributor to the *St. James's Gazette* from 1881–90. An edited version of this obituary appeared in the *St. James's Budget* (21 Jan. 1898), accompanied by two Tenniel illustrations ("Alice growing taller" and "The Mad Hatter and the March Hare cramming the Dormouse into the Teapot").

as he was only in his sixty-sixth year, he was by no means an old man as age goes in these long-lived days.

Clergyman and Mathematician.

Mr. Dodgson, who was born in 1833 at Croft, near Darlington, was the son of the Archdeacon of Richmond. He was educated at Christ Church, Oxford, and distinguished himself in the schools, taking a first class in Mathematical and a second in Classical Moderations, and a first in the Final Mathematical School, and a third in "Literæ Humaniores." He became a Senior Student of Christ Church in 1861, and in the same year mathematical lecturer, a post which he continued to fill for twenty years. In 1861 he was also ordained; but never proceeded to priests' orders, owing to an impediment in his speech. He began his literary career in 1860 by the publication of *A Syllabus of Plane Algebraical Geometry*, which was followed the next year by *The Formulæ of Plane Trigonometry. A Guide to the Mathematical Student in Reading, Reviewing, and Writing Examples* made its appearance in 1864.

Some of His Propositions in *The Game of Logic.*

His knowledge of higher mathematics, which was extensive and profound, enabled him to discharge the duties of Mathematical Lecturer from 1855 to 1881, and at one time it was generally expected that he would be raised to the University chair of that science. He was the author of several bright and amusing mathematical pamphlets, one of which—*Euclid and the Game of Logic*—well deserves to be republished. In one of those writings he parodies the propositions of Euclid with inimitable wit. Take, for instance, the unanswerable, self-evident, and lamentable proposition: "A discussion may be raised on any point at any distance from that point." Or this other, which should be carefully borne in mind by all public orators of all countries at the present day: "A plain superficial speaker is one who, any two points being given, will lie evenly between these points."

Alice in Wonderland.

The Adventures of Alice in Wonderland appeared in 1865. The book was originally written to amuse one of Dean Liddell's daughters. The author was an intimate friend of the Dean and Mrs. Liddell, and took infinite pleasure in the society of their little girls. It was in order to beguile her hours of playtime that these diverting fancies were woven for one of the children. The quiet quaintness and pedantic precision which characterise the wild whimsicalities of the *dramatis personæ* of this fairyland of nonsense are unrivalled, and the verses, "You are old, Father William," are as popular as the most finished productions of our classics. The story rapidly circulated throughout the English-speaking world, and was translated into all the languages of Europe. Seven years later (1872) appeared the continuation of *Alice*, under the title *Through the Looking-Glass*, and although only a sequel, it was scarcely inferior to its predecessor, and was almost equally appreciated. These two works are masterpieces of the exuberant fancy, graceful style, and poetic genius of Lewis Carroll. Nothing that he wrote before or after—and numerous ingenious literary efforts have been since associated with his name—comes at all near to these. The verses entitled "The Jabberwock," which are composed in a language as artificial as "Volapuk" or "Esperanto," are none the less familiar to and understanded of the nursery.

A Little Surprise for the Queen.

The success of *Alice* was never in doubt, and the story is current, though we cannot vouch for its authenticity, that the Queen herself on reading it was so much delighted that she commanded the author to send his next work to Windsor. He did so, and Her Majesty was almost as bewildered as Alice on finding that it consisted of *An Elementary Treatise on Determinants*!

His Love of Children.

Mr. Dodgson's beautiful suite of rooms in the north-west corner of Wolsey's great quadrangle, looking over St. Aldgate's, were at one time a veritable children's paradise. Never did rooms contain so many cupboards, and never did cupboards contain such endless stores of fascinating things. These rooms in Christ Church were a real Wonderland for many a real child. From Mr. Dodgson's cupboards musical boxes, mechanical performing bears, picture-books innumerable, toys of every description, came forth in bewildering abundance before the child's astonished eyes; no wonder, then, that in childish years a day spent with "Lewis Carroll" was like a glimpse into a veritable El Dorado of innocent delights. Nor were Mr. Dodgson's little friends all of one class. He took for many years a deep interest in homes and hospitals for sick children, and no inconsiderable portion of the profits of his books was devoted to these works of charity. Mr. Dodgson had beneath his twinkling humour a deep fund of seriousness and religious earnestness. He used at one time to issue a fly-leaf to "every child who loves Alice," in which he sought to mingle grave thoughts of charity and faith with fun and frolic.

As a Play-goer.

He developed an amazing love for the comedy stage. Into that singular phase of his career there is no time at present to enter. He was the friend and philosopher of the original "Little Lord Fauntleroy;"[1] his dis-

[1] Dodgson saw Vera Beringer play the title role in *The Real Little Lord Fauntleroy*, Frances Hodgson Burnett's own dramatization of her 1886 novel, on 18 June 1888 and made friends with her that same day.

appearances from Christ Church had nearly always reference to a visit to some play; and, excepting his love for children—a love which had been warmly reciprocated, and is now saddened by the sense of loss—the play was, perhaps, the only thing he could in later years develop any enthusiasm for.

Some Little Eccentricities.

Mr. Dodgson was particular about his appearance. He hated to be introduced as "the author of Alice," and his short-sightedness was sometimes exaggerated with a purpose. Of recent years he revolted against the discount system, and followed Mr. Ruskin in insisting (with some modification) on net prices. He would not advertise his books (and, indeed, they hardly needed it) in the ordinary way, but he sometimes inserted announcements on his own account of an original character in the leading journals.[1] He never wore an overcoat in his life, and continued his long "constitutionals to the end."

His Odd Ways.

A writer in the *Daily Chronicle* once asked Mr. Dodgson for some biographical facts and for a photograph. Unhappily, the reply was "in the negative, with a half-brick," as the humourist has said. Others, no doubt, have had similar disappointments. The friends who knew him best—those who did not desire an "imitated autograph" were content to know that he lived his extraordinary life by himself and largely for himself. One who has been the friend of all the world could scarcely be the enemy of a biographer. It is a fact, nevertheless, that he could be almost insolent to such inquirers; and as for the young ladies who sought his autograph, it afforded him much pleasure to contemplate the posibility of a reunion of them at which a comparison of the signatures might be made.

An Essay In Etymology.

Mr. Dodgson was to the last an uncommonly pleasing and popular man, whose conversation was often suggestive and always amusing. Despite the duties of his calling, which he discharged with scrupulous care, he had numerous acquaintances in the theatrical world of London, who respected his learning, humoured his amiable eccentricities, and keenly appreciated his wit. The sayings attributed to him at Oxford would fill an entertaining volume of Carrolliana. Among other things, his "ety-

---

[1] For instance, Dodgson placed such advertisements in *The Times* and the *Daily News* in 1893 and 1894 recalling copies of *Looking-Glass* which had been poorly printed.

mology of the bell" is still quoted with relish by scholars. There was a provisional belfry at Christ Church College, which was familiarly known to Oxonians of the time as "the meat safe." Mr. Dodgson, undertaking to explain this epithet etymologically, split up the word belfry into two parts—the French word belle and the German word frei (free). Then he went to work as follows:

Belle=beautiful=comely=meet (meat);

Frei=free=secure=safe.

Result: "Meat-safe."

"As Thou Didst the Egyptian."

One of his best repartees was elicited by an Oxford student driving a tandem. This candidate for academical distinction had the social honor of driving the late Khedive of Egypt through the ancient city of Oxford, and the misfortune to have "spilled" his Highness. On the following day the youth, driving the same tandem, overtook Mr. Dodgson, who looked somewhat tired. "May I give you a lift?" he asked jauntily. To whom the humourist: "Wilt thou slay me, as thou didst the Egyptian yesterday?"

The Mathematician and the Fabulist.

Mr. Dodgson was a very odd and original mixture of a prim mathematician and a happy fabulist. Here is an attempt at a list of the books not above mentioned, which he published. It will be seen that many of them are of a very severe cast. In 1869 he published *Phantasmagoria and other poems*; in 1870 *Songs from 'Alice's Adventures in Wonderland,'*; in 1871 *Facts, Figures, and Fancies relating to the Elections to the Hebdomadal Council. Euclid Book V., Proved Algebraically*, made its appearance in 1874, to be followed two years later by another example of the author's versatility in *The Hunting of the Snark, an Agony in Eight Fits.* His subsequent works include *Doublets: a Word Puzzle*, 1879; *Rhyme? and Reason?* 1893; *A Tangled Tale*, 1885; *Alice's Adventures Underground*, 1886; *The Game of Logic*, 1887; *Curiosa Mathematica, Part I.—A New Theory of Parallels*, 1888; and *Symbolic Logic*, 1896. His latest work, *Sylvie and Bruno*, must have come as a disappointment to his friends. Whereas his earlier works transport children to a fairy realm of fun and frolic, in which not even the transient shadow of a moral flits across the perpetual sunshine of delight, a large portion of *Sylvie and Bruno* is taken up with what must be termed sermons and lessons. The author of *Sylvie and Bruno* is not the Lewis Carroll who composed *Alice in Wonderland*, but the Rev. Charles Lutwidge Dodgson, clergyman.

St. Nicholas (New York)[1] March 1898

This was that brave adventurer
Upon an unknown sea,
Who found the far, fair Wonderland—
His galleon by an eager band
Of little children featly manned,
All laughing out in glee.

Far, far away his vessel sailed,
Throughout a single night,
Until it reached that magic shore
No man had ever seen before;
The children's land forevermore
He gave them as their right.

And since that voyage venturesome
On every night and day
That pilot with a shipful new
Of happy children for his crew—
Of grown-up folk a favored few—
Has sailed the Wonder-Way.

And if upon to-morrow's ship
No Pilot should appear,
So many children everywhere
Have learned from him the thoroughfare
To Wonderland, they still will dare
To sail without a fear.

But oh! their little hearts will ache,
And oh! their eyes will dim;
And, as the ship sails mile by mile,
Each child will sit a little while,
And, thinking, will forget to smile—
For sailing without him.

---

[1] This poem, "Lewis Carroll," was signed by Abbie Farwell Brown (1871–1927), poet and author of children's books. As a teenager, Brown was coeditor of *The Jabberwock*

St. Nicholas (New York) March 1898

Miss Abbie Farwell Brown's poem on page 375 of the present number of *St. Nicholas* will have special interest to all readers who are familiar with *Alice in Wonderland*, and other books by Lewis Carroll, whose real name was Charles Lutwidge Dodgson.

Surely the two names represent two very different men: one, an English clergyman and for twenty-six years lecturer on mathematics at Christ Church college, Oxford, author of many volumes on his special subject, honored by his fellows, and peer of England's distinguished men of learning—the other, the merriest, quaintest of story-tellers, whose name is known everywhere by thousands of English-speaking readers, young and old.

Dear Lewis Carroll! bright, delightful, preposterous, gifted with a wonderful imagination, yet so modest and shy that he told almost in a whisper the story of *Alice in Wonderland* that made all England shake with laughter. *Through the Looking Glass* was issued, one may say, as a profound public secret, and he sent bashfully, secretly, into the world *The Hunting of the Snark*, one of the jolliest, most audacious little books ever written. To the very last, their author shrank from any mention of these delightful works as though they were an offense to his other self—that profound, serious scholar, whose discourses, lectures, and bewildering mathematical problems would have sent Lewis Carroll scampering into the first hiding-place in Wonderland.

One day, about ten years ago, the writer accompanied by a friend, stood in the large somber study of Mr. Dodgson, in Christ Church college. As we entered the room we could see at first only the heavy table with piles of pamphlets and great leather volumes upon it; and, around it books, books, and more books—ponderous and worm-eaten (we felt sure they were worm-eaten). Then a slight but stately figure rose from somewhere in the shadow behind an open volume. With a nervous little cough he came forward, bowing so stiffly, and with so slight a show of surprised delight at our coming, that for an instant we felt awed and subdued. Next we realized that he was bashful, actually timid! But, later, his gentle voice, his grave, kindly eyes, and his gracious words satisfied

---

(the magazine of the Boston Latin School), to which Dodgson contributed a poem and wrote some letters. See also below p. 169.

us that he really was the man whose genial letter, crossing the seas, had made us regard him as a friend. Among other subjects, he spoke pleasantly of *St. Nicholas*, and of young Americans; but when, in the most delicate manner possible, we ventured an allusion to *Alice in Wonderland*, instantly his brow clouded; there came a sudden reserve of manner, a silent, dignified "Spare me!" that changed the conversation, like magic. We realized that we were speaking to the Rev. Charles L. Dodgson, of Oxford University, England, and not to Lewis Carroll, author of *Alice's Adventures*—and *Through the Looking-Glass*. Yes, it was the great thinker—the scholar and mathematician—whom we were visiting, and who, when our pleasant call was over, solemnly bade us good-by. And there was no thought of the "March Hare" or of "Father William" in our minds as we went softly down the dark stairway, out into the sunlight of the grand old quadrangle of Christ Church college.

And yet—do you know?—we felt inwardly sure that in some brisk Jabberwocky way Mr. Dodgson secretly loved little Alice and her adventures, and at heart was not sorry, after all, that he had written them.

St Pancras Gazette (London)[1]

What glad dreams that sound awakens
Of the past, thy passing bell,
How we mourn thee, friend of childhood,
Loved so long and loved so well.

"They lived happy ever after,"
Thus the fairy tales were told
Round the ingle-side. What day-dreams
They inspired in days of old.

How the happy little faces
Of the fairy folk would beam,
Sunlit was each glowing ember
In the ruddy fire-light's gleam.

[1] "This poem, "In Loving Memory of Lewis Carroll," was signed "W. E. Heard"

Castles rose by glistening fountains,
Never-fading bloom were there,
Fairy footsteps cross'd the mountains.
Ah! how beautiful they were.

Now we see the path before us
Lighted by no fairy flowers,
And the faith of happy childhood
Never can again be ours.

Now we have grown older, wiser,
Is such wisdom greater gain,
Though we tread life's sunny places,
Or may know its keenest pain?

Still the joys of happy childhood
Cling to us in riper years,
Still we see thy fairy faces
Shining through a mist of tears.

The Saturday Review (London) January 22, 1898

There is a tendency in our contented acceptance of a received classic to forget what force of originality it had at the time of its appearance, to forget, that is to say, how largely it appealed to its first readers as an imaginative discovery—a view which perforce gets weakened when a succeeding generation has become infected, as it were, by the fineness of its quality. There is, no doubt, this difference to-day in our appreciation of Lewis Carroll's work, that we accept it with minds less arrested by its novelty than by its familiarity and abounding humour. More than any other English writer Carroll has attracted to himself with equal force the crabbed age and the youth of his generation. For that, mere cleverness cannot account: to have taken such root, to have become so naturalised, his work must have struck a note of peculiar appropriateness to his day: it was as it were a microbe that fell upon a constitution happily prepared for its inroads. Yet his rendering of fairy-tale was a departure from tradition.

Of all true fairy-tale this may be said: it has never been "written down" to the intelligence of childish readers: rather it has been the meeting-point of interests, the medium through which the child can experience the thrill of big-hearted adventure, and the grown man the romantic sense and undisturbed belief in poetic justice which are the birthright of childhood. For centuries the child has warmed its imagination upon the folk-tales which once gave contentment to its less simple-minded elders, and do still have an interest for them more or less vague. But up to the latter half of this century the workings of a child's mind were of little interest to its elders. Now, however, an astonishing change has come over the dream of mankind; and to-day the child is the grown-up man's fairy-book. Almost by instinct a wise mind stoops to look into that under-world so close at hand, so stored, perhaps, with the wisdom of the ancients; and so a new ground for fairyland has been reclaimed from the prose of existence. To minds still sensitive to chivalry and romance, the old fairy-tale remains true; but a far larger world of believers has been gathered to the imaginative world of childhood, and arrested, almost studiously, to a contemplation of its laws.

It was through this change of outlook, and as almost its first portent, that Lewis Carroll's fairy-books struck the light. Without condescension, without any "putting back of bright intelligence," rather with all the resources of wit and humour, and a sub-pathetic sense of the ridiculous, Carroll, a man of grave reserve and fine intelligence, turned round upon all tradition to find a new fairyland ready-made, fresh-sprung from humanity's new interest in the things of childhood.

The seriousness of his work, and its safety, come from the fact that it is as much fairy-tale for elders as for children; and the highest interest of these two time-divided classes is the interest they have in common. Lewis Carroll's ability to evoke this common interest was the measure of his success; the material for it was his great discovery. This tendency to bring grown-up minds where a little child shall lead them showed a further advance twenty years after *Alice in Wonderland* had appeared, in Stevenson's *Child's Garden of Verse*, where, almost too much, it is the elder's mind that is appealed to, and the youngster stands somewhat out of it through being analysed to excess. Yet even that charming failure at combined interests indicates the road along which men's absorption in youth has progressed since then.

It may seem strained, in the face of the mirth-moving spirit of the *Wonderland* and *Looking-glass* tales, to claim for them so much purpose and seriousness; but few can feel that their quaint wit and baffling logic were but the outcome of a mere sense of fun. Wherever there is beauty in human things there must also be pathos, something intent if not wistful:

and in many of Carroll's tender extravagances there is beauty of a high order. Take, as an instance, the Fawn that has forgotten its name and accompanies Alice fearlessly till the wood is cleared; then, crying, "I am a fawn, and you—you are a human child!" springs startled away. So, also, the whole episode of the White Knight, of so much wisdom gone foolishly astray, has about it almost the same charm as Millais' "Sir Isumbras," now delighting all who make Burlington House their mild wintry resort.[1] There is satire, too, light and stingless, but satire none the less, conveyed under a cajoling disguise of parody; and some may be able to see in the "aged, aged man a-sitting on a gate" a not distant reference to Wordsworth's "leach-gatherer,"[2] and in the figure of the White Knight, with his horse-like face, some likeness to the poet himself.[3]

Lewis Carroll was too much a scholar and a recluse in temperament not to find many things contrary to his taste in the world from which, almost with acerbity, he guarded his private personality; its slipshod speech and illogical ways of thought received many a passing rap from his wit; and the desire for interrogation and argument which his heroine Alice, charming type of her age, carries to the length of foolishness, is resolutely put down with a finality which leaves the incorrigible little offender breathless.

> What did they draw?" said Alice, quite forgetting her promise. "Treacle," said the Dormouse, without considering at all this time. "But I don't understand. Where did they draw the treacle from?" "You can draw water out of a water-well," said the Hatter: "so I should think you could draw treacle out of a treacle-well—eh, stupid?" "But they were in the well," Alice said to the Dormouse. "Of course they were," said the Dormouse,—"Well in." . . . . "And they drew all manner of things—

---

[1] "A Dream of the Past—Sir Isumbras at the Ford" (1857), a painting by John Everett Millais first displayed at Burlington House, London. The painting shows an old armor-clad knight on a horse with two children. See Russell Ash, *Sir John Everett Millais* (London: Pavillion, 1996), plate 19. Dodgson saw the painting on 1 July 1857 and wrote that it was "remarkably ugly. . . . The knight is good, though with an expression like an honest old gardener."

[2] The White Knight's Song from *Looking-Glass* is a travesty on "Resolution and Independence," by William Wordsworth (1770–1850), a poem which features the leech-gatherer mentioned above. See Gardner, *The Annotated Alice*.

[3] In fact, Tenniel's illustration of the White Knight is thought to be a self-portrait.

> everything that begins with an M,—" "Why with an M?" said Alice. "Why not?" said the March Hare. Alice was silent.

So, too, every effort she makes to moralise on a situation gets baffled,—

> "I like the Walrus best," said Alice, "because, you see, he was a little sorry for the poor oysters." "He ate more than the Carpenter, though," said Tweedledee. "You see he held his handkerchief in front, so that the Carpenter couldn't count how many he took: contrariwise." "That was mean!" Alice said, indignantly. "Then I like the Carpenter best—if he didn't eat so many as the Walrus." "But he ate as many as he could get," said Tweedledum. This was a puzzler.

Alice, indeed, moves through her wonder-world with much of the modern spirit, which has now and then to be wholesomely repressed. It is also noticeable, that in "Wonderland" to her alone come magic attributes making her alternate from gigantic to microscopic proportions. And, as giant, she fares badly, in the good old way, at the hands of the natives. They, on the other hand, behave like ordinary unreasonable human beings: the positions, therefore, of the traditional fairy-tale are reversed. But even against her might the moral order of Wonderland stand; and she is expelled for bad behaviour.

If Wonderland was her school, the Looking-glass world was her college. She enters as a freshman, having learned by now to do without unreasonable habits of growth; but her inquisitiveness is unabated, and her moralisings have positively increased. And so in the end she is sent down for her moralisings, when she takes the law, and the august person of the Red Queen, into her own hands. And because of her importunacy in the ways of Providence we shall never know what would have happened at that gorgeous feast where the candles went up like sky-rockets, and the guests and the viands changed places.

Is not all this a warning against that spirit of the age which will ask too many meanings, and try to import too much reason into the Edens still marvellously left to us? It seems as if Lewis Carroll was sent to poke fun with unexampled urbanity at the new age which has sprung upon us, and yet, from many grave and tender reasons, to welcome it.

The Scottish Guardian (Edinburgh)[1] January 28, 1898

The late Lewis Carroll (if Lewis Carroll can ever be called *the late*) had a just horror of the interviewer. He was a great deal too modest to blow his own trumpet, or to let anyone else blow it for him. A few personal reminiscences, however, by one of his comtemporaries, who had special opportunities of observing him, will be no breach of confidence, and may prove interesting to the readers of *The Scottish Guardian*, numbers of whom must have come under the spell of his almost unique genius. It was my fortune to be matriculated at the same time with him, and to have some conversation with him as we walked down together to the Vice-Chancellor's, under the guardianship of Osborne Gordon,[2] another Christ Church character.

The Vice-Chancellor at that time was Dr. Plumptre,[3] a remarkably tall and remarkably thin personage. He it was who gave rise to the mathematical conundrum, "Why is the Vice-Chancellor like a line? Because he is length without breadth"; while his successor, Dr. Cotton,[4] a singularly small man, was defined as a point, which "hath neither parts nor magnitude"; and his predecessor, Dr. Symonds,[5] a big man all round, was described as a cube, or "that which hath length, and breadth, and thickness." Oxford men of the period will remember the picture of the three Vice-Chancellors in Wyatt's window in connection with the conundrum which hit off their physical and their reputed mental characteristics.

On coming into residence I had the advantage of reviving an acquaintance thus begun, being associated with Charles Lutwidge Dodgson in the same "mess." Members of "The House,"[6] whose memories can carry

---

[1] This articles, titled "Personal Recollection of the Author of *Alice in Wonderland*," was signed "G. J. C.-B." and written by George James Cowley-Brown (1832–1924), Chancellor and Canon of Edinburgh Cathedral. He matriculated at Christ Church on 23 May 1850, took his B.A. in 1854 and his M.A. in 1857.

[2] Senior Student of Christ Church, 1834–61, censor 1847–59. Matriculated 1832 at age 18. Became rector of Easthampstead, Berkshire in 1861 until his death in 1883.

[3] Frederick Charles Plumtre, matriculated at University College in 1813, Master from 1836 until his death in 1870. Served as Vice-Chancellor from 1848–51.

[4] Richard Lynch Cotton, matriculated in 1811 at Worcester College, Fellow from 1816–38, provost of Worcester from 1839 until his death in 1880. Served as Vice-Chancellor from 1852–56.

[5] Benjamin Parsons Symons, matriculated 1802 at Wadham College, Fellow from 1812–31, Warden of Wadham, 1831–71, died 1878. Served as Vice-Chancellor 1844–48.

[6] i.e. Christ Church, from its Latin name "*Aedes Christi*," *aedes* being Latin for "house."

them back to the period, will call to mind the batches of half-a-dozen undergraduates who dined together at the different tables in the hall, and the disgraceful way the dinner at that time was served. The hour then was five. Though the spoons and forks were silver—some of them very old, the gift of former members of The House—the plates and dishes were pewter. The joint was pushed from one to another, each man hacking off his own portion, and rising from the table without waiting for one another, without even waiting for the ancient Latin grace, said by one of the junior students in response to the invitation of the senior scout (a stout man with a squeaky voice) always in the same drawling tone, "Gra-ace, Gen–tlemen, plea-ease." For the reform of all this sumptuary slovenliness, and other customs "more honoured in the breach than in the observance,"[1] we are indebted to the late Dean (father of the original Alice) who departed this life only a few days after the friend who had taken his child for a prototype.

In Dodgson's "mess" were young Philip Pusey,[2] whose crippled frame enshrined a pure soul and a cultivated intellect; the late Rev. G. G. Woodhouse;[3] the present rector of St. John's, Edinburgh,[4] and others equally unknown to fame. The features, however, of one of our number are immortalised in "The Hatter," who figures in the "Mad Tea-party" in *Alice in Wonderland.* Among our contemporaries, bachelor or undergraduate, were the late Lord Lothian,[5] a man of the very highest promise; the late Lord Carnarvon,[6] and his successor in the vice-royalty of Ireland, Lord Cowper;[7] Lord Harrowby, the late Duke of Newcastle,[8] the present Bishop of Gibralter, Lord Justice Vaughn Williams, Sir Henry Longley,[9]

---

[1] *Hamlet*, Act I, scene iv.

[2] Philip Edward Pusey, oldest son of Edward Bouverie Pusey. Matriculated at Christ Church October, 1850, B.A. 1854, M.A. 1857, died 1880 at the age of 49.

[3] George Girdlestone Woodhouse, matriculated Christ Church October 1850, B.A. 1854, perpetual curate of Upper Gornal, 1861–67.

[4] Here the author refers to himself.

[5] William Schomberg Robert Kerr, matriculated Christ Church 1850, B.A. 1854,

[6] Henry Howard Molyneaux Herbert, matriculated Christ Church 1849, B.A. 1853, D.C.L. 1859, high steward Oxford 1859, Keble College council 1871–82, 4th Earl of Carnarvon and father of the 5th Earl who was the patron of Howard Carter, discoverer of the tomb of King Tutankhamen.

[7] Francis Thomas de Grey Cowper, matriculated Christ Church in June 1852, B.A. and M.A. 1861, Lord-Lieutenant of Ireland 1880–82.

[8] Henry Pelham Alexander Pelham Clinton, matriculated Christ Church 1852.

[9] Son of Charles Thomas Longley, Bishop of Ripon and close friend of Charles Dodgson, Sr. Henry Longley was at Rugby in the same house with Charles Dodgson (the younger), matriculated at Christ Church in 1852 and became a barrister-at-law in 1860.

Canon Jelf,[1] Mr. Talbot,[2] one of the present members for the University; the present Chancellor of the Exchequer, Sir Michael Hicks-Beach;[3] Mr. Oswald,[4] afterwards curate of St. John's, Edinburgh; Malcolm of Poltalloch,[5] and many others more or less known to fame. We all, however, I may safely say, sat in the same hall and some of us even at the same table with Dodgson without discovering (perhaps from our own want of it) the wit, the peculiar humour, that was in him. We looked upon him as a rising mathemetician, nothing more. He seldom spoke, and the slight impediment in his speech was not conducive to conversation. One day, however, long after we had taken our degrees, he asked me to go for a walk with him. On our return, he said that if I would come to his rooms, he would show me some drawings by Tenniel. They were in pencil, on small squares of paper, and were the originals of the illustrations in *Alice.* He then confided to me that he was writing a book. One morning, not long after, he, like Byron, woke "to find himself famous."[6]

I have spoken of our "mess." Those whose memories can reach back to Lord Derby's installation[7] may remember a scene which occurred at it on that occasion. At the banquet in Hall the high table was, in the opinion of the undergraduates, unduly honoured at the expense of the other tables. Dish after dish was carried past us. It was repletion at the top, inanition at the bottom. The dons were stuffed, the rest of us were starving. They were at their last course before we had been well supplied with our first. To redress or protest against this inequality, one audacious youth stepped out of his place near the door, as one of the *scouts* was entering, bearing aloft a colossal structure of barley-sugar in the shape of a castle, the *chef-d'œuvre* of the *manciple*, on its way to the high table. This he seized, and set it down on our table, where it collapsed, and was

---

[1] George Edward Jelf, matriculated Christ Church 1852, B.A. 1856, M.A. 1859, Canon of St. Alban's 1878–80, Canon of Rochester 1880.

[2] John Gilbert Talbot, matriculated Christ Church 1853, B.A. 1858, M.A. 1860, Member of Parliament for West Kent 1868–78, first elected MP for Oxford University 1878.

[3] Sir Michael Edward Hicks-Beach, matriculated Christ Church 1855, B.A. 1858, M.A. 1861. First served as Chancellor of the Exchequer 1885–86. First served as MP (East Gloucestershire) 1864–85.

[4] Henry Murray Oswald, matriculated Christ Church 1851, B.A. 1855, M.A. 1860.

[5] John Wingfield Malcolm, matriculated Christ Church 1850, B.A. 1856. M.A. 1865.

[6] George Noel Gordon, Lord Byron (1788–1824), entry in Memoranda after the publication of the first two cantos of *Childe Harold's Pilgrimage*, as quoted in Thomas Moore's 1830 *Life of Byron.*

[7] Edward Geoffrey Smith Stanley (1799–1869), 14th Earl of Derby, was installed as Chancellor of Oxford University in 1852 and served in that post until 1869.

shivered in pieces. The proceeding so tickled poor Pusey's fancy, that he broke out into those screams of inextinguishable laughter which the stone-deaf, who cannot realise the sound of their own voices, sometimes send forth. This was followed by a dead silence, the Dean and the dons and their distinguished guests standing up and straining their eyes towards the quarter from which the unearthly sounds were issuing.

It was Dr. Pusey, the father of this excellent but afflicted young man, who gave Dodgson his studentship. In those days the studentships of Christ Church (a species of fellowship) were, with the exception of the Westminster ones, pieces of patronage which the Dean and Canons divided between them. Dodgson, however, though the son of one of Dr. Pusey's *collaborateurs* in *Library of the Fathers*,[1] was, if I may use the word, no "Puseyite."[2] I may note here his telling me of a rendering by his father which he thought aptly imitated the alliteration of the original, the phrase, *a cœno ad cœlum*[3] (somewhere, I think, in Tertullian), which he had rendered "from the sty to the sky." Dodgson was an unaffectly religious man, as witness his *Easter Greeting*, with its now pathetic reference to his "hope to look back upon" what he had written "without shame and sorrow . . . when *my* turn comes to walk through the valley of the shadows." He was no party man, neither was he a "Puritan," as one of the party church newspapers last week thought proper to describe him. His shyness and the consciousness of his stammer did not at the time I am speaking of prevent his occasionally helping one of the city clergy; and I remember his enlisting my help also, during a short residence in Oxford, on behalf of one of the hard-working chaplains who held one of these churches. My subsequent interviews with him, our lot being now cast in different countries, were "few and far between," but on occasional visits to Oxford he was always gald to see me. I recall one of these occasions on which he asked me to come to his rooms after Hall, and on ascertaining that I had a daughter of the age of *Alice*, he sent her a beautiful copy of the book with her name inscribed in it "from the author." I ventured to ask him once what made him take this *nom de plume*. He said, "You see my name is Charles, and as Carolus is the Latin for it, I thought Carroll would do very well. My name is also Lutwidge, and Lewis comes near enough to it."

---

[1] Archdeacon Dodgson's contribution to the series of translations of early Christian writings undertaken by Pusey was a translation of Tertullian published in 1842.

[2] Pusey was a strong supporter of the Oxford movement, whereas Dodgson disliked any tendency towards ritualism and any movement of the English Church towards Rome.

[3] From the mire to the sky.

My last conversation with him was on the occasion of the *Gaudy*[1] last June. I sat by him during the customary speech in the chapter-house, and I remember his telling me that it fell to him once to make a speech there on *Hakluyt's Voyages*, Hakluyt having been an old Westminster student of the House.[2] He dined with us in Hall, but I saw him no more, having to leave that night, little thinking how soon he was to be taken from the society he had helped to sweeten. There were no signs of senility about him. The wine of his fancy ran clear to the last. "*Quando ullum invenient parem?*"[3]

The Sketch (London)[4]
January 26, 1898

Nobody except a few Oxford people knew the Rev. Charles Lutwidge Dodgson, who has died at the age of sixty-five. Everybody has heard of "Lewis Carroll" and of *Alice in Wonderland.* I have in my possession a letter written by this unique humorist to a child friend which illustrates his love of children. The child has been staying in apartments with friends, and has upbraided the creator of "Alice" with not coming to see her, insinuating that he did not perhaps deem her acquaintances sufficiently attractive. To this charge "Lewis Carroll" replies—

> I don't want any more acquaintances for myself. It isn't pride! Please don't think it! It is that I am old, and very tired of Society, and of interchanging commonplaces with strangers. That was the sort of thing that came into my mind. And my conclusion was I would much rather avoid it all and put up at a hotel.
>
> Let me add that, supposing the lodgings had been those of

---

[1] An annual dinner for past undergraduates of Christ Church.

[2] It was customary at Gaudy for a Christ Church B.A. to deliver a speech about a former distinguished member of the college. Dodgson delivered his speech on Richard Hakluyt (1552?–1616) on 31 May 1856, though Dodgson's diary records that he gave the speech in Hall, not in the chapter-house. Dodgson's speech was published in 1974 in *The Lewis Carroll Circular*, number 2.

[3] When shall we find his like again?" Horace, *Odes* I.24.8.

[4] This obituary is accompanied by the photograph of Dodgson by W. Shawcross of Guildford.

your friend (let us say), the Archbishop of Canterbury,[1] and that he would probably be "looked in" upon by a few other Bishops, my conclusion would have been the same—to go to the hotel, and avoid the whole lot of them. And now do, dear child, be friends as we were before.

The Spectator (London) January 22, 1898

One cannot truly say that by the death of Lewis Carroll the gaiety of nurseries has been eclipsed, for it is practically certain that even had he lived another ten or twenty years we should have had no further additions to the *Alice* cycle. The magician's wand was already broken, and though his spells retained and will retain their power as long as the English language lasts, there was no hope of another *Alice* or even of another *Snark*. *Sylvie and Bruno* proved that. Though one or two of the conversations and some of the snatches of song had something of the old enchantment, the wizard's circle was not complete. The charm was not fully "wound up," and could not hold and compel the spirits of laughter and delight. But to say this is to seem ungracious and ungrateful. In spite of the failure in *Sylvie and Bruno*, Lewis Carroll accomplished more in the way of giving absolute and unalloyed pleasure to young and old than any man of his generation. The two *Alices*, the *Snark*, and the poems collected in *Rhyme and Reason*, many of which have never quite had their due, made the world imcomparably richer in the region where it most needs enrichment,—the region of pure merriment. It is comparatively easy to set men laughing if one depends upon satire, upon burlesque, or upon the humourous incongruities of real life. Lewis Carroll never relied upon these pedestrian means for securing the delight of his audience. His genius carried always with it a suggestion of clear, and yet soft, laughing sunshine. He never made us laugh at anything, but always with him and his knights and queens and heroes of the nursery rhymes. He seemed, as it were, to fetch his laughter from some limpid and elemental source where it had never been profaned with human tears of earthly contacts. It ripples and dances in the sunbeam of his fancy with-

[1] Unfortunately this letter is not dated, so we cannot be certain who the Archbishop of Canterbury was at the time. If the letter was written between 1862 and 1882, the Archbishop would have been an acquaintance of Dodgson (Archbishop Longley, former Bishop of Ripon, 1862–68, or Archbishop Tait, former headmaster of Rugby School, 1868–82).

out any reminder of the fact—though fact it doubtless is—that side by side with the river of laughter runs the river of tears, and that most of those who dip their pitchers mingle the two streams. It is a commonplace of criticism that the great humourists have in them always something of sighs as well as of smiles. Lewis Carroll never for a moment makes us incline to sadness, even in the midst of mirth. There is nothing pathetic even in the White Knight or the White Queen, and not an emotion is stirred by these "blessed creatures," the Beaver, the Snark, or the Jubjub, which is not compounded of mere pleasure. Curiously enough, this could not quite be said of Lear's delightful creations. There was often an infinitesimal touch of sadness in his fantasies.

It is very difficult to analyse the nature of Lewis Carroll's work. *Alice in Wonderland*, and the marvelous sequel, which was even better than the original, *Alice through the Looking-Glass*, took not only England, but the world, by storm. Both books were almost at once translated into every European language, and reached in a very few years the position of classics. No one would ever dream of talking about a book called *Alice in Wonderland*. *Alice* is quoted by writers and speakers exactly as one quotes Scott, or Dickens, or Thackeray. What is the quality which makes Lewis Carroll's books so extraordinarily attractive? We believe that attraction to be due to the fact that they possess in the highest possible degree the literary quality of charm. They win their way into our hearts by their easy grace and fluent eloquence. If one considers the matter for a moment, it is impossible not to notice that this special gift of charm is comparatively rare with humourists. Even *The Rose and the Ring*[1] is almost without it, and, as a rule, Thackeray in his humourous passages is devoid of charm. He has plenty of charm when he is serious or pathetic, but not when he is funny. In the older dispensation the comic masque was not thought to require charm. Rather it was held to be best without it. But charm was, as we have said, the dominant characteristic of Lewis Carroll's work. His style, with its perfect rhythm, its perfect taste, and its perfect clearness and easiness of tone, is a most alluring guide into that wonderland behind the looking-glass "where, entered once, all curious pleasures come" to welcome and delight us. This is seen most obviously in his poetry. The songs in the two *Alices* haunt the memory so potently because of their charm. But of course Lewis Carroll had a great deal more than charm. He had a power over the fantastic so complete and so masterly that he could give it full rein without ever allowing it to become

---

[1] Popular Christmas book (1855) by William Makepeace Thackeray (1811–63).

grotesque or burlesque. Even in his most extravagant moments he never committed any of these wild enormities upon which lesser minds rely. To say this is only to say in other words that there was always a kindly humanity in all Lewis Carroll's creations. Indeed he was a true son of Oxford and of "humane letters." The Walrus, the Mock-turtle, the March Hare, and all the other monsters, the playing cards and chessmen, have each of them a something which warms our hearts. As the children say, they are all "such dears," and we love them as we laugh. What added very greatly to the charm of Lewis Carroll's writing was doubtless the double power by which he was at once elusive and allusive. One can never quite run his ideas to earth, yet there are always ideas there,—ideas just vanishing round the corner like the tail of this white rabbit, and ideas, too, which have some queer and quaint half-intended allusion to something, though what that something is we cannot quite tell. As Alice says herself, when she has heard the poem of the "Jabberwock" recited, "Somehow it seems to fill my head with ideas—only I don't exactly know what they are." Here is the authentic art of the nonsense-writer. To seem allusive and then be elusive. Had Lewis Carroll really been allusive he would have been a mere satirist. As it is, when we think we have caught the allusion to something social or political, it vanishes or becomes something else, just as things keep vanishing and changing in the shop kept by the Old Sheep. Possibly it was this very quality of elusive allusion that in a rougher and ruder way constituted the charm of the Elizabethan jesters. They poked their fun at everything and every one, but they were never satirists and never unkindly. Their wit played like a lambent flame, but never scorched and never settled on one object. As a proof of what we mean by Lewis Carroll's power of elusiveness combined with allusiveness take the delightful conversation with the Red and White Queens after Alice has become a Queen:—

> "I don't deny things with my hands," Alice objected.
>
> "Nobody said you did," said the Red Queen. "I said you couldn't if you tried."
>
> "She's in that state of mind," said the White Queen, "that she wants to deny something—only she doesn't know what to deny!"
>
> "A nasty, viscious temper," the Red Queen remarked; and then there was an uncomfortable silence for a minute or two. The Red Queen broke the silence by saying to the White Queen, "I invite you to Alice's dinner-party this afternoon."
>
> The White Queen smiled feebly, and said "And I invite you."
>
> "I didn't know I was to have a party at all," said Alice; "but if

there is to be one, I think I ought to invite the guests."

"We gave you the opportunity of doing it," the Red Queen remarked: "but I daresay you've not had many lessons in manners yet?"

"Manners are not taught in lessons," said Alice. "Lessons teach you to do sums, and things of that sort."

"Can you do Addition?" the White Queen asked. "What's one and one and one and one and one and one and one and one and one and one?"

"I don't know," said Alice. "I lost count."

"She ca'n't do Addition," the Red Queen interrupted. "Can you do Subtraction? Take nine from eight."

"Nine from eight I ca'n't, you know," Alice replied very readily: "but—"

"She ca'n't do Subtraction," said the White Queen. "Can you do Division?" Divide a loaf by a knife—what's the answer to that?"

"I suppose—" Alice was beginning, but the Red Queen answered for her.

"Bread-and-butter, of course. Try another Subtraction sum. Take a bone from a dog: what remains?"

Alice considered. "The bone wouldn't remain, of course, if I took it—and the dog wouldn't remain: it would come to bite me—and I'm sure I shouldn't remain!"

"Then you think nothing would remain?" said the Red Queen.

"I think that's the answer."

"Wrong as usual," said the Red Queen: "the dog's temper would remain."

"But I don't see how—"

"Why, look here!" the Red Queen cried. "The dog would lose its temper, wouldn't it?"

"Perhaps it would," Alice replied cautiously.

"Then if the dog went away, its temper would remain!" the Queen exclaimed triumphantly.

Alice said, as gravely as she could, "They might go different ways." But she couldn't help thinking to herself "What dreadful nonsense we are talking!"

"She ca'n't do sums a bit!" the Queens said together, with great emphasis.

"Can you do sums?" Alice said, turning suddenly on the White Queen, for she didn't like being found fault with so much.

> The Queen gasped and shut here eyes. "I can do Addition," she said, "if you give me time—but I ca'n't do Subtraction under any circumstances."

It is impossible to doubt that the elusiveness of this dialogue, its deliciously hide-and-seek character, constitutes a great part of the charm. But equally delightful is the inchoate allusiveness. Think of the possible applications of "She's in that state of mind that she wants to deny something, only she does not know what to deny!" It looks like a telling allusion to half the controversies, political, theological, and social, that are now engaging mankind. Yet, of course, Lewis Carroll was not thinking in the least of anything satirical. So with the enchanting remark a little further on: "A little kindness—and putting her hair in papers—would do wonders for her." There are moments when that seems like a most pointed reference to some one we all know, but as we clasp the allusion it melts into thin air. Another example of what we mean is afforded by Lewis Carroll's poems. Looked at in certain lights, they seem like conscious attempts to parody great contemporary poets. The moment, however, we settle down to noting which is the poet meant the resemblance utterly vanishes. The allusion has become elusive. The poem recited by the White Queen, for example, looks almost as if it were meant as a parody of Browning:—

> "First, the fish must be caught."
> That is easy: a baby, I think, could have caught it.
> "Next the fish must be bought."
> That is easy: a penny I think, would have bought it.
> "Now cook me the fish!"
> That is easy, and will not take more than a minute.
> "Let it lie in a dish!"
> That is easy, because it already is in it.

As soon, however, as we begin to note the resemblances we see that nothing but pure fun was intended, and that even the good-humoured satire of parody was not meant. Again, "The Aged, Aged Man" might by a careless reader be taken as a parody of Wordsworth, but, in truth, it is nothing of the kind. But we have said enough, perhaps more than enough, of Lewis Carroll's books, for work woven in the merry eyes of children is best enjoyed when least criticised. In Lewis Carroll's case it is imperative that we should be like Wordsworth's poet, he who was,

> "Contented if he might enjoy

The things that others understand."[1]

The *Alices* and the *Snark* can only be enjoyed fully by those who will approach them with the heart of a little child. And what child wants to know why and how Lewis Carroll's books delight him or her? Let us be content to know that to this generation has been given a rill of laughter which does not fail, in which no bitter has been mixed with the sweet, and at which all may drink, whether old or young.

Standard (London) January 15, 1898

The Rev. Charles L. Dodgson ("Lewis Carroll"), the author of *Alice in Wonderland*, died yesterday at the Chestnuts, Guildford, the residence of his sisters, in his sixty-sixth year. Mr. Charles Lutwidge Dodgson was born in 1832, educated at Christ Church, Oxford, and took his degree of B.A. in 1854, and that of M.A. in 1857. He was ordained deacon by the Bishop of Oxford in 1861, but did not proceed to priest's orders. He was appointed mathematical lecturer of Christ Church in 1855, and occupied that position until 1881. During that period he wrote and published a large number of works upon mathematical subjects, including *Euclid and his Modern Rivals, Curiosa Mathematica, a New Theory of Parallels*, a *Syllabus of Plane Algebraical Geometry*, &c. But the works by which he gained fame, and the affection of all children, were those by which he enlivened his drier studies. Such were *Alice's Adventures in Wonderland*, the first of the interesting series, which appeared in 1865*; Through the Looking-glass, and What Alice Saw There*, which followed an *Elementary Treatise on Determinants. Euclid, Book 5 Proved Algebraically*, was succeeded by *The Hunting of the Snark*, and the last of the unique series of children's stories was *Alice's Adventures Underground*, published in 1886.

Sunday Times (London) January 16, 1898

Lewis Carroll—to give the writer the name by which he endeared himself to many thousands of readers—will be mourned with heartfelt sorrow by those whom he has delighted with the brightness of

[1] Wordsworth, "A Poet's Epitaph," 55.

his humour and the freshness of his fancy. Who has not read *Through the Looking Glass, Alice in Wonderland, Sylvie and Bruno*, and their companions—that series of irresistibly and whole-heartedly comical story-books? If the gaiety of nations could be eclipsed, it would be at the passing away of such a writer as the late Mr. Dodgson. Children especially, when they hear the sad news, will grieve over the loss of a real friend.

Surrey Advertiser (Guildford) January 15, 1898

We sincerely regret to have to announce the death of the Rev. Charles Lutwidge Dodgson, the celebrated author of *Alice in Wonderland*, which occurred yesterday (Friday) afternoon at the Chestnuts, Quarry-street, the residence of his sisters, the Misses Dodgson. The deceased gentleman was the eldest son of the late Archdeacon Dodgson, rector of Croft, Yorkshire. He was educated at Rugby under Dr. Tait, and from there he went to Christ Church, Oxford. Here he obtained first-class mathematical honours in 1854, and received the B.A. degree in the same year. In 1861 he was ordained deacon in the diocese of Oxford. For twenty-six years he was mathematical lecturer at Christ Church. A large number of books on various branches of mathematics have come from his pen, amongst which may be mentioned *Euclid and his Modern Rivals* and *Curiosa Mathematica*. The deceased gentleman was more widely known by his nom de plume of "Lewis Carroll," and under this name he has perhaps given to the children of England, whom he loved so well, the most fascinating story ever written, for *The Adventures of Alice in Wonderland* has deservedly become the classic of modern children's books. Amongst many other books which have come from his pen may be mentioned *Alice through a looking glass*, *Sylvie and Bruno*, *The Hunting of the Snark*.

Surrey Advertiser (Guildford)[1] *January 22, 1898*

The death of the Rev. C. L. Dodgson, better known as Lewis Carroll, author of *Alice in Wonderland*, which we announced in last Saturday's *Surrey Advertiser*, has been received with profound regret

[1] Much of the material in this article is quoted or paraphrased from other sources,

throughout the whole country. Several daily papers on Saturday devoted a good deal of their space to an account of the life of the celebrated author. As a writer, says one of our contemporaries, Lewis Carroll belonged to no officially recognised category; he broke entirely fresh ground—was at once a humourist, a poet, a satirist, a mathematician, a polished man of the world, and a clergyman. If ever wit and humour were born with a man they were with Lewis Carroll, whose clever repartee, sudden sallies, and comical comparisons took the form of winged words, and circulated from Land's End to John o' Groats. It is not necessary in these columns to recapitulate what has been said so much better by our contemporaries. Unstinted praise, praise of a kind which falls to the lot of few modern authors, has been bestowed upon that most delightful book, *Alice in Wonderland*, a book that will remain with us for all time. The tale was written, it is said, to amuse the weary hours of a sick child,[1] the principal character, Alice, being modelled on one of the daughters of the Dean of Christ Church, and it was with considerable reluctance that the author consented to its publication. It immediately leaped into popularity, and it is said that the Queen was so charmed with it that she ordered that all the books by the same author should be sent to her, and was surprised and dismayed to find that the greater number of them were dry mathematical pamphlets.

After *Alice in Wonderland*, the sequel, *Alice through the Looking Glass*, was the most popular. *The Hunting of the Snark* also met with a good reception. Among Mr. Dodgson's other works, some of which we mentioned last week, were *Euclid and the Game of Logic*, *Phantasmagoria*, a collection of parodies and poems, *Doublets*, *Euclid and his Modern Rivals*, *A Tangled Tale*, *The Game of Logic*, and *Sylvie and Bruno*.

There are many anecdotes quoted of Mr. Dodgson's quaint humour, smart repartees, and witty parodies.

The following is one of his happy parodies on the propositions of Euclid: "A discussion may be raised on any point at any distance from that point," and another, "A plain superficial speaker is one who, any two points being given, will lie evenly between those points."

Among his witty sayings at Oxford the following "Etymology of the Belfry" is quoted. There was a provisional belfry at Christ Church College, which was familiarly known to Oxonians of the time as "the meat safe." Mr. Dodgson, undertaking to explain this epithet etymologically,

---

especially *The Daily Telegraph* (15 Jan. 1898), see above pp. 37–40 for additional footnotes.

[1] Alice was not sick when the story was first told (see below pp. 6n, 82n).

split up the word belfry into two parts—the French word belle and the German word frei (free). Then he went to work as follows: "The word 'Belfry' is derived from the French 'bel'—beautiful, becoming, meet (meat)—and from the German 'frei'—free, unfettered, secure, safe. Thus the word is strictly equivalent to 'meat-safe' to which the new belfry bears a resemblance so perfect as almost to amount to coincidence."

One of the smartest repartees attributed to him is the following: One of the students had the honour of driving the late Khedive of Egypt through the ancient City of Oxford and the misfortune to have "spilled" his Highness. On the following day the youth, driving the same tandem, overtook Mr. Dodgson, who looked somewhat tired. "May I give you a lift?" he asked, jauntily. To whom the humorist: "Wilt thou slay me as thou didst the Egyptian yesterday?"

For many years Mr. Dodgson was an amateur photographer, and amused himself by taking his little friends in all sorts of odd and fanciful costumes, till his albums became filled with Japanese boys and girls, beggar-maids in picturesque tatters, or Joans of Arc in glittering armour.

It was not easy to realise that Mr. Dodgson was 65. His hair was indeed iron grey; but his alert figure and his bodily vigour seemed to betoken a younger man. He never wore an overcoat in his life, and he continued his long "constitutionals" to the end.

Mr. Dodgson's beautiful suite of rooms in the north-west corner of Wolsey's great quadrangle, looking over St. Aldgate's, were at one time a veritable children's paradise. Never did rooms contain so many cupboards, and never did cupboards contain such endless stores of fascinating things. These rooms in Christ Church were a real Wonderland for many a real child. From Mr. Dodgson's cupboards musical boxes, mechanical performing bears, picture-books innumerable, toys of every description, came forth in bewildering abundance before the child's astonished eyes; no wonder, then, that in childish years a day spent with "Lewis Carroll" was like a glimpse into a veritable El Dorado of innocent delights.

Mr. Dodgson was once immensely amused at the candid criticism of a child friend of his. He asked her if she had read *Alice* and the *Looking Glass*. "Oh, yes," she replied readily, "I've read both of them. And I think" (this more slowly and thoughtfully)—"I think *Through the Looking Glass* is more stupid than *Alice's Adventures*. Don't you think so?"

Referring to the death of Mr. Dodgson at Christ Church, Oxford, on Sunday morning, Canon Sanday, Margaret Professor of Divinity, said all that made the individual, the infinite play of fancy and the subtle undercurrents of serious and chastened thought must needs be lost to them:

they went with him whose they were to inhabit another sphere than ours. The world would think of "Lewis Carroll" as one who opened out a new vein in literature, a new and a delightful vein which added at once mirth and refinement to life. His books were wholly without spot or stain in the midst of so much that was the reverse of stainless, and were all seasoned with such a wholesome salt of humour as won for them a wide popularity. Might they not say that from their courts at Christ Church there had flowed into the literature of their own time a rill bright and sparkling, healthgiving, and purifying wherever its waters extended? Perhaps there were few happier spots in the English literature of the century than its books for children, and how large a share of this was due to him who was gone! For this all England was grateful, nay, more, all readers of the English tongue.

THE FUNERAL

The funeral of the deceased took place on Wednesday morning at Guildford, and was remarkable for the simplicity which, in conformity with the deceased's written instructions, the arrangements were carried out. The coffin, which was entirely hidden by the profusion of choice floral tributes, was conveyed from the Chestnuts, the residence of the deceased's sisters, on a hand bier to St. Mary's Church, and there was only one mourning coach, the majority of the mourners walking to the church, and subsequently to the cemetery.[1] The body was met at the entrance to the church by the Dean of Christ Church, Oxford (the Very Rev Dr Paget), the Rev Canon Grant (rector), and the Revs A. Kingston and E. C. Kirwan (curates). The mourners were the Rev S. H. Dodgson and Mr W. L. Dodgson (brothers),[2] Miss Elizabeth Dodgson, Miss Caroline Dodgson, and Miss Margaret Dodgson (sisters),[3] Mr C. H. W. Dodgson (nephew),[4] and Mr J. H. D. Dodgson (cousin),[5] and Mrs Wilcox, of Winchester (cousin).[6] Among others present at the service were the

---

[1] The walk to the church was a short one—down the hill through Castle Arch and two hundred yards or so down Castle Street. The walk to the cemetery, on the other hand, was a long steep climb.

[2] Rev. Skeffington Hume Dodgson (1836–1919) and Wilfred Longley Dodgson (1833–1914).

[3] Elizabeth Lucy Dodgson (1830–1916), Caroline Hume Dodgson, Margaret Anna Ashley Dodgson (1841–1915).

[4] Charles Hassard Wilfrid Dodgson (1876–1941), older son of Dodgson's brother Wilfred.

[5] James Hume Dodgson (1845–1912), fourth son of Hassard Hume Dodgson, brother to Dodgson's father.

[6] Probably Frances Rooney Wilcox (Fanny) (1832–1919), wife of Dodgson's cousin William Edward Wilcox.

Rev T. J. Prout, the Rev T. B. Strong, Mr A. G. V. Harcourt, and Mr A. Hassall (all of Christ Church),[1] Mr G. L. Craik[2] (representing Messrs Macmillan), Dr Gabb,[3] the Rev H. R. Ware, Dr Clark, Messrs G. C. Williamson, F. W. Guilding, G. C. Wheat, R. Mason, J. Hooke, and J. H. Wiles. The service in the church was conducted by Canon Grant, assisted by the Rev A. Kingston, and the Dean of Christ Church read the lesson. The hymns sung by the choir were "The saints of God! their conflict past," and "My God, my Father, while I stray," and as the body was borne from the church the organist (Mr F. J. Hooke) played "O, rest in the Lord." At the cemetery the committal sentences were read by the Dean of Christ Church, and the choir sang the hymn, "Peace, perfect peace." The coffin, which was of plain oak, bore the inscription:—"Charles Lutwidge Dodgson, born January 27, 1832; died January 14, 1898."

The wreaths and crosses were from his loving brothers and sisters, his affectionate nephews—Stuart and Bertram Collingwood and Wilfred Dodgson, Mrs and Miss Wilcox (Winchester), in affectionate remembrance; Mr Fletcher Lutwidge,[4] in affectionate remembrance; the Dean of Christ Church, Oxford, and Mrs Paget in affectionate remembrance; the members of the Common-room, Christ Church, Oxford,[5] Mr J. H. Arkwright,[6] Christ Church, Oxford, Mr and Mrs Robert E. Baynes[7] (Oxford), in affectionate memory; Mrs Reginald Hargreaves,[8] (Oxford), Mrs and the Misses Lucy,[9] in affec-

---

[1] All Students of Christ Church. Thomas Jones Prout, B.A. 1846, M.A. 1848, vicar of Binsey (1857–91); Thomas Banks Strong, B.A. 1883, M.A. 1886; Augustus Charles Vernon Harcourt, B.A. (Balliol) 1858, M.A. (Christ Church) 1861, natural science examiner, delegate of University Museum; Arthur Hassall, B.A. 1877, M.A. 1880 (Trinity), examiner in history, member of the Hebdomadal Council.

[2] George Lillie Craik, friend and, through Dodgson's publisher, business associate.

[3] James Percy Alwyne Gabb (1853–1934), cared for Dodgson during his final illness.

[4] Charles Robert Fletcher Lutwidge (1835–1907). Dodgson's cousin, the son of his mother's brother Charles Henry Lutwidge.

[5] Dodgson curated the Senior Common Room, Christ Church, from 1882–92, see Wakeling, *The Oxford Pamphlets, Leaflets, and Circulars of Charles Lutwidge Dodgson.*

[6] John Hungerford Arkwright, matriculated Christ Church 1852, B.A. & M.A. 1864, lord of the manor of Hampton.

[7] Robert Edward Baynes (1849–1921), B.A. 1872 (Wadham), M.A. 1875 (Christ Church), Senior Student at Christ Church and Lee's Reader in Physics.

[8] Alice Pleasance Hargreaves, née Alice Pleasance Liddell, daughter of Henry George Liddell, Dean of Christ Church. The child friend for whom *Alice* was written.

[9] Mrs. Alice Jennings Lucy (wife of William Lucy) and her daughters Edith Elizabeth (1871–1907) and Catherine Susanna (Katie) (1870–1934). Dodgson met Edith when she was a student at the Oxford High School and he was giving logic lessons and subse-

tionate remembrance; Miss Isa Bowman,[1] in loving memory; Miss Enid Stevens[2] and Miss Margaret Mayhew[3] in loving remembrance from two of his child friends; Miss Weld,[4] Oxford; Mrs and Miss Harding, Guildford, with kind regards and much sympathy; the Misses "Mary" and "Cuckoo" Morritt, Miss Hartshorne, and Miss C. Hartshorne, with deep sympathy and regrets; Mrs E. Southwell,[5] in affectionate remembrance; Mrs and Miss Blakemore (bouquet);[6] Mrs Thicknesse, "Edith and Dolly,[7] with love;" Mrs Walters and family,[8] with most affectionate remembrance; the servants at the Chestnuts; Mrs George G. Woodhouse, "widow of 'C. L. D's.' first friend at Christ Church, Oxford,"[9] and "In loving memory of the sweetest soul that ever looked with human eyes, from E. G. T."[10]

The funeral arrangements were carried out by Messrs. Pimm and Sons, Guildford.

---

quently befriended the family. Both girls were married by 1898, but still signed their wreath "The Misses Lucy."

[1] Isabella (Isa) Bowman (1874–1958), one of Dodgson's closest and most enduring child-friends. She played Alice in the 1888 revival of Savile Clarke's London stage adaptation of *Alice in Wonderland* (see above p. 9n) and once visited Dodgson at Eastbourne for five weeks. Author of *The Story of Lewis Carroll* (1899).

[2] Enid Gertrude Stevens (1882–1960), child-friend. Dodgson dedicated *Sylvie and Bruno Concluded* to Enid.

[3] Margaret Dorethea Mayhew (1883–1971), child-friend, daughter of Anthony Mayhew, chaplain and Hebrew lecturer at Wadham College, Oxford.

[4] Probably Agnes Grace Weld, child friend and niece of Tennyson whom Dodgson had photographed as Red Riding Hood in 1857 and who appears in Dodgson's diary as late as 1895.

[5] Mrs. Edmund Martin Southwell, née Ethel Florence Barclay (1864–1947), daughter of Henry Alexander Barclay, who was an undergraduate at Christ Church with Dodgson. Dodgson frequently visited the Barclays in Brighton and sometimes took Ethel on excursions to London.

[6] Mrs. Villiers Blakemore (née Sarah Elizabeth King) and her daughter Edith Rose ("Dolly") Blakemore (1872–1947). Edith was a child-friend whom Dodgson had met in Eastbourne.

[7] Mrs. Thicknesse was the wife of Francis Henry Thicknesse, but as Agnes Beatrice Jane Argles ("Dolly") (1857–1944) had been a child friend of Dodgson, as was her sister Edith Margaret Argles (1853–1935). Edith was later Vice-Principal of Lady Margaret Hall, Oxford, and Dodgson gave a series of logic lectures to students there.

[8] The Walters family were friends of Dodgson at Guildford. Mrs. Henry Littlejohn Master Walters (née Harriet Mitchell) and her husband had five children.

[9] Mrs. George Girdlestone Woodhouse (née Emmeline Benson) (1838?–1921). When George Woodhouse (1835–97) died, Dodgson wrote to Mrs. Woodhouse, "Of all the friends I made at Christ Church, your husband was the very *first* who spoke to me—across the dinner-table in Hall."

[10] Emily Gertrude Thomson (see above p. 53n).

Surrey Advertiser (Guildford) January 29, 1898

Sir. Will you kindly disclaim the joke about the Khedive as having been made by Lewis Carroll. This joke is mentioned in your paper of January 22nd as attributed to him, but he could not have made it, as he most strongly objected to any witticisms connected with words of Scripture.

I am, Sir, One of His Mourning Relatives, 4 Park Crescent, Brighton, Jan. 27th, 1898.[1]

Sydney Morning Herald[2] January 17, 1898

The death is announced of Lewis Carroll (the Rev. Charles Dodgson), the author *of Alice's Adventures in Wonderland*, and other works.

The Rev. Charles Dodgson, who was better known under the pseudonym of Lewis Carroll, was senior student of Christ Church, Oxford. He was an eminent authority in mathematical studies, and was the author of *A New Theory of Parallels* and *Symbolic Logic.* Lewis Carroll published *Alice's Adventures in Wonderland* in 1865, *Through the Looking-Glass and What Alice Found There* in 1871, and *Alice's Adventures Underground* in 1886.

The Times (London)[3] January 15, 1898

We regret to announce the death of the Rev. Charles Lutwidge Dodgson, better known as "Lewis Carroll," the delightful author of *Alice in Wonderland*, and other books of an exquisitely whimsical humour. He died yesterday at The Chestnuts, Guildford, the residence of

---

[1] Stanley Godman in the article "Lewis Carroll's Sister: Henrietta Dodgson" (*Notes and Queries,* New Series, vol. 5, no. 1 [Jan. 1958]) identifies Carroll's sister, Henrietta Harrington Dodgson (1843–1922), as the author of this letter.

[2] The same obituary appeared in the *Sydney Daily Telegraph* on 17 Jan. 1898.

[3] This obituary was reprinted almost exactly in the *Warrington Guardian* (19 Jan. 1898) and in slightly edited form in the *South Western World and Clapham Junction*

his sisters, in his 66th year. He was educated at Christ Church, Oxford, and distinguished himself in the Schools, taking a first class in Mathematical and a second in Classical Moderations, and a first in the Final Mathematical School, and a third in *Literae Humaniores*. He became a Senior Student of Christ Church in 1861 and in the same year, mathematical lecturer, a post which he continued to fill for 20 years. In 1861 he was also ordained. He began his literary career in 1860 by the publication of *A Syllabus of Plane Algebraical Geometry*, which was followed the next year by *The Formulæ of Plane Trigonometry. A Guide to the Mathematical Student in Reading, Reviewing, and Writing Examples* made its appearance in 1864, and in 1865 *The Adventures of Alice in Wonderland* burst upon an astonished world. Few would have imagined that the quiet, reserved mathematician, a bachelor, who all his life was remarkable for his shyness and dislike of publicity, possessed the qualities necessary to produce a work which has stood the test of more than 30 years, and still captivates young and old alike by its quaint and original genius. This was the first, or one of the first, of those entertaining books, since become numerous, which afford almost equal enjoyment to boys and girls and to those children of a larger growth who, although years have rubbed off the bloom of their youthful illusions, yet preserve their love of innocent laughter and nonsense. *Alice in Wonderland* was originally written to amuse one of Dean Liddell's daughters. The author was an intimate friend of the Dean and Mrs. Liddell, and took infinite pleasure in the society of their little girls. It was in order to beguile her hours of playtime that these diverting fancies were woven for one of the children. The success of the book was never in doubt, and the story is current, though we cannot vouch for its authenticity, that the Queen herself on reading it was so much delighted that she commanded the author to send his next work to Windsor. He did so, and her Majesty was almost as bewildered as Alice on finding that it consisted of *An Elementary Treatise on Determinants*! It is curious to notice how frequently *Alice in Wonderland* is quoted in reference to public affairs, as well as to the ordinary matters of every day life. Hardly a week passes without the employment of its whimsicalities to point a moral or adorn a tale, and only yesterday a letter from a correspondent was published in *The Times* in which the Dreyfus-Esterhazy case was paralleled, with an aptness which

---

Gazette for Putney, Wandsworth, Battersea, Clapham, Streatham, Balham, Tooting, Wimbledon and District (22 Jan. 1898). Two additional items which appeared in *The Times* were reprinted in their entirety in *The Oxford Chronicle* (see above pp. 113–14).

was really surprising, from Lewis Carroll's immortal story. Some years ago Alice made her appearance on the London stage in a graceful dramatic version of the book, and delighted the children home for the Christmas holidays, for whose especial benefit ocular demonstration was given of her surprising experiences. In 1869 Lewis Carroll published *Phantasmagoria and other poems*; in 1870 *Songs from 'Alice's Adventures in Wonderland,'* in 1871 *Through the Looking-Glass and What Alice Found There,*—a continuation which obtained almost the success of the original work—and in the same year *Facts, Figures, and Fancies relating to the Elections to the Hebdomadal Council. Euclid, Book V., Proved Algebraically*, made its appearance in 1874, to be followed two years later by another example of the author's versatility in *The Hunting of the Snark, an Agony in Eight Fits*. His subsequent works include *Doublets: a Word Puzzle*, 1879; *Rhyme? and Reason?* 1883; *A Tangled Tale*, 1885; *Alice's Adventures Underground*, 1886; *The Game of Logic*, 1887; *Curiosa Mathematica, Part I—A New Theory of Parallels*, 1888; and *Symbolic Logic*, 1896.

Although, as is abundantly evident from this list of his works, Lewis Carroll was a serious and a hardworking mathematician, there can be no doubt that his chief title to fame will always rest on those jeux d'esprit which have won for him so secure a place in the affections of readers for whom mathematics are, as a rule, the reverse of attractive. In many a home and many a schoolroom there will be genuine sorrow to-day when it is announced that the author of *Alice in Wonderland* and *Through the Looking-Glass* has passed away.

The Westminster Gazette (London)[1] January 15, 1898

Not "the late Lewis Carroll," though, unhappily, we are compelled to speak this morning of the late Mr. Dodgson. There are suggestions in various places this morning that Mr. Dodgson was somewhat jealous of the commanding reputation of "Lewis Carroll." It was his

[1] In addition to the pieces listed here, *The Westminster Gazette* also printed the anecdote about Dodgson going into a room full of adults on all fours (under the title "Another Lewis Carroll Story" on 15 Jan.), see above p. 88; a summary of Dr. Paget's Christ Church sermon (under the title "Dean Liddell and Lewis Carroll" on 24 Jan. 1898) see below pp. 176–80 for complete text; and an abbreviated account of the funeral under the title "Funeral of Lewis Carroll" on 19 Jan.

dearest wish to be known as the author of "serious works," such as treatises on the higher mathematics. The list of his works, set out by his biographers this morning, is pathetic evidence of the fact. Lewis Carroll on Alice is followed by Dodgson on Determinants; the *Hunting of the Snark* takes its place with *A New Theory of Parallels*. Mr. Dodgson was indefatigable, while Lewis Carroll was intermittent. Towards the end, according to some of his friends, Mr. Dodgson was apt to brood over a "life mis-spent" by Lewis Carroll. There is nothing strange in this. It is the last infirmity of genius to suppose that its real achievements lie in a line which the world refuses to recognise, and the humorous, in particular, are always remorseful about the lost opportunities of "serious work."

We must leave Mr. Dodgson's claims to mathematicians, and be content to say that Lewis Carroll is one of the benefactors of these times. No one, since Charles Lamb, has hit upon a vein of humour so absolutely original and so full of meaning for young and old. For himself, he would have none of the solemn meanings which ingenious people read into his innocent books. They were what they were, and he meant nothing more by them than was set down in black and white. All the same, he was a master of the most delicate double entendre, whether he owned to it or not. He delighted the child and was a deep joy to the grown-up. *Alice in Wonderland* and *Alice through the Looking Glass* have in this respect the quality of all fine literature. They adapt themselves by some magic to your age and your experience. You read them on the nursery floor, and rejoiced in the breathless proceedings of the White Rabbit, while the Hatter, the March Hare, the Cheshire Cat, the Duchess, and the Mock Turtle made up the population of a fairy-land whom you somehow felt to be nearer to you and more human than the impossibly beautiful or dreadfully wicked people who haunted the pages of other writers thought suitable to your age. Then as you grew up, you find that though other nursery books became foolish, these grew with you, expanding like the telescopic Alice. There were things in them which suited almost every situation in life, and at each reading you seemed to discover something new in them, or some fresh application of the old things. Only yesterday, somebody discovered for us that the trial of the Knave of Hearts furnishes the most delightful satire imaginable on the Dreyfus business. We quoted it yesterday, and a correspondent writes to us this morning that it served precisely the same purpose and with equal effect eight or nine years ago when the Parnell letter was the subject of a State trial in this country. We turn at random to the other volume and open at the White Knight. How immensely significant he has become, since we first made his acquaintance in the nursery! Hear him lecturing on the art of riding and falling off between each sentence. Look at his equipment, so delightfully represented in Sir John Tenniel's picture with something ("my own invention") for every conceivable emergency. He has a box to keep clothes and sandwiches in, carried upside down "so, that the rain can't get in," the only drawback to which is that the things can get out. He has an excellent beehive fastened to the saddle "in hopes some bees may make a nest in it," and a mousetrap, in case some mice should appear. The Knight's philosophy is that "it's as well to be provided for everything," and the horse has anklets round his feet "to guard against the bites of sharks." We are ashamed to recall anything so familiar, but surely he is an immortal type of a considerable portion of the human race, even when he goes on talking head downwards at the bottom of a ditch from which only the soles of his feet emerge. The applications are a hundredfold. Similarly we turn the page again and find the delightful assertion, "What I say three times is true,"[1] as if to hit off in a sentence the most confirmed habit of the human mind in its treatment of evidence.

Here we must pause. You have only to turn the page to multiply instances. The whimsicality of it, the delightful muddle of ideas, as though one hemisphere of the brain had somehow got mixed with the other, are the speciality of the writer, and no one has imitated him successfully. Heaven forbid that anyone should call them "serious" works, but it is at least a very serious achievement to have contributed thus much to the delight of men, women, and children.

The Westminster Gazette (London)[2] January 15, 1898

By a singular coincidence yesterday's leading article in the *Westminster Gazette* contained a long extract from *Alice in Wonderland*;[3] this morning we have to chronicle with much regret the death of "Lewis Carroll," otherwise the Rev. C. L. Dodgson. He was staying with his sisters at Guildford, where he passed away yesterday in his sixty-sixth year. The sad event will cause sorrow all the world over, for probably no writer has produced books so universally popular with children and grown-up people as *Alice in Wonderland* and *Through the Looking-Glass*.

---

[1] The Bellman says "What I tell you three times is true," in *The Hunting of the Snark*.
[2] This article was signed "H. A. S."
[3] Quoted from *The Times* (see below pp.174–5 for text and context).

The Identity of "Alice."

A quarter of a century ago the Rev. Charles Lutwidge Dodgson was a fellow and mathematical tutor of Christ Church, and the daughters of the former Dean, Dr. Liddell, an attractive group of little girls, occupied a prominent place among his small friends. There is some doubt as to which of them the stories of "Alice" are actually dedicated to, but tradition commonly points to Mrs. Reginald Hargreaves, Alice Liddell as she then was. That the name is the same is certainly an argument in favour of this view. Some, however, maintain that the dedication was in reality addressed to the next sister, Edith, whose untimely death upon the eve of her marriage with Mr. Harcourt, of Newnham [*sic*] Park, was the cause of very deep grief and regret to all who knew her.[1] Apart from the tragedy of her death at such a moment, she was universally beloved for her own sake, and was as gifted as she was in every way charming. In any case, however, though the dedication poem of *Alice Through the Looking-Glass* is very obviously intended for one only of his little friends,[2] that at the beginning of *Alice in Wonderland*, and also the one at the close of the looking-glass adventures, would seem to be rather addressed to three of the sisters together. On reading these graceful little verses one might easily infer that the first idea of this fantastic dream-tale had flitted across the author's brain upon a certain drowsy summer afternoon as he drifted in a boat along the Oxford river, while the three little girls pulled manfully at the oars, and at the same time begged for a fairy story from their ever-willing friend.

Mr. Dodgson and the Children.

Anyone who reads the adventures of Alice in later years (said a recent appreciation) will appreciate the marvellous knowledge of child nature which Mr. Dodgson possesses. No doubt he started with a large share of that most indefinable gift for understanding and sympathising with the young; but though he has himself always remained a bachelor, his experience of children has been none the less extremely wide. In Oxford alone, among his university acquaintances, there was not a house which owned children, and particularly little girls, which he did not frequent, and where a warm welcome was not always ready for him. His appearance, generally at odd hours, and flavoured with that, to children, delightful element of the unexpected, was a signal for lesson books to be discarded and for a general rush to the drawing-room, with a happy dis-

---

[1] But Edith's death did not occur until some years later (see above, p. 21n).

[2] It is an acrostic on the name Alice Pleasance Liddell.

regard of unbrushed hair or soiled pinafores. The children knew perfectly well that this visitor came to see them, and that they were not only to be seen, but heard. The parents also acquiesced in the arrangement. Mr. Dodgson was always admitted to be unlike other people, and his popularity with the children insured him perfect liberty to pursue the unconventionality of his ways. He had seldom been in the house three minutes, before, with pencil and paper, he had embarked the elder of his young friends upon some elaborate puzzle or conundrum, while jokes of all sorts and weird fairy tales were reserved for the little ones who clustered around him. One of the greatest treats that the children could have was to go to tea with Mr. Dodgson, in his quaint rooms in the Tom Quad, at Christ Church, where games and entertainments without number were always ready for them, and where the hours passed only too quickly. Better still, perhaps, to be snatched from the dreary routine of a walk with nurse or governess, and borne off, regardless of remonstrance on the part of these worthies, to the mysterious gloom of a half-darkened room, to be photographed.[1] Mr. Dodgson was very fond of taking portraits at a time when amateur photography was scarcely as finished as it is at present, and many little groups of boys and girls, faded now almost out of recognition,[2] are still preserved as mementoes of these childish treats.

His Mathematical Works.

As the *Daily News*[3] well points out, Mr. Dodgson and Lewis Carroll were in authorship, and to some extent in life also, two very different beings. Mr. Dodgson was the don, and writer on mathematical subjects; Lewis Carroll was the author of charming books for children. Mr. Dodgson graduated at Christ Church, Oxford, in 1854—having taken a first class in mathematics and a third in Lit. Hum. In the following year he was appointed mathematical lecturer to the College, a post which he continued to occupy until 1881. In 1858 he was elected to a Senior Studentship, and this he continued to hold until the end. In 1861 he was ordained. His earliest, as well as his latest, publications, were all mathematical—*A Syllabus of Plane Algebraic Geometry* (1860), *The Formulæ of Plane Trigonometry* (1861), *A Guide to the Mathematical*

[1] In fact, wet-plate photography required a great deal of light and Dodgson took most of his photographs outdoors or in on his rooftop studio (see above p. 57).

[2] Actually, prints made from the glass negatives of the wet plate process seem much less susceptible to fading than many other photographs. Even today most of Dodgson's extant photographs are quite clear.

[3] For the complete footnoted text of the *Daily News* article, see above, pp. 33–6.

Student (1864). The first of his children's books was published in 1865. Other mathematical works followed, of which the more important were: *An Elementary Treatise on Determinants* (1867), *Euclid, book V., proved Algebraically* (1874), and *Curiosa Mathematica: A New Theory of Parallels* (1888).

And the Inimitable Lighter Volumes.

Alice's Adventures in Wonderland was originally written in MS. for the amusement of his girl friends at Oxford, and a facsimile of it in that form was published many years afterwards. It first saw the light in 1865, and at once became a popular favourite. "So much clever and yet genuine fun," said one of the reviewers, anticipating the general verdict, "has never before been found within the compass of one volume."[1] Edition followed edition, and by this time *Alice* must have appeared in some fifty editions. The first edition is now very scarce, and is much sought after by collectors—a distinction which it partly owes to Sir John Tenniel's graceful and amusing illustrations.[2] Nor is the fame of *Alice* restricted to the English-speaking folk; for it has been translated both into German and into French. The sequel *Through the Looking-Glass, and What Alice Found There*, published in 1871, was equally successful, and it has run through as many editions as the original *Alice*. Mr. Dodgson used to be fond of repeating the deliciously naive remark of a child-friend, whom he asked, after an acquaintance of two or three days, if she had read *Alice* and the *Looking-glass*. "Oh yes," she replied readily, "I've read both of them. And I think" (this more slowly and thoughtfully) "I think *Through the Looking-glass* is more stupid than *Alice's Adventures*. Don't you think so?" Next to the *Alice* volumes, Lewis Carroll made his principal hit with *The Hunting of the Snark: An Agony, in Eight Fits*. This appeared in 1876, and had to be reprinted several times in the same year. It was subsequently embodied in a larger volume called *Rhyme and Reason* (1883). Of his remaining books, *Sylvie and Bruno* (1889), with its *Conclusion* (1893) is the best known. *Alice's* and *The Hunting of the Snark* have by this time become household words.

---

[1] On 22 December 1866 a reviewer for *The Spectator*, noting that *Alice* had reached a fifth thousand, wrote "so much clever and yet genuine fun in the letter-press, and so much grace and humour in the illustrations have never before been found within the same compass."

[2] That the collectible value of the first edition is due in part to Tenniel cannot be argued, for it was Tenniel who wrote to Dodgson expressing disappointment with the poor quality of the printing, which eventually lead to Dodgson's scrapping the edition and creating a great rarity (see also p. 90n).

His Rooms a Real Wonderland.

Mr. Dodgson's beautiful suite of rooms in the north-west corner of Wolsey's great quadrangle, looking over St. Aldgate's, were at one time a veritable children's paradise. Never did rooms contain so many cupboards, and never did cupboards contain such endless stores of fascinating things. These rooms in Christ Church were a real Wonderland for many a real child. From Mr. Dodgson's cupboards musical boxes, mechanical performing bears, picture-books innumerable, toys of every description, came forth in bewildering abundance before the child's astonished eyes. Nor were Mr. Dodgson's little friends all of one class. He took for many years a deep interest in Homes and Hospitals for Sick Children, and no inconsiderable portion of the profits of his books was devoted to these works of charity.

The Serious Side of His Character.

Mr. Dodgson had beneath his twinkling humour a deep fund of seriousness and religious earnestness. As the *Daily News* reminds us, he used at one time to issue a fly-leaf to "every child who loves Alice," in which he sought to mingle grave thoughts of charity and faith with fun and frolic.

> God does not mean us to divide life into two halves—to wear a grave face on Sunday, and to think it out of place to even so much as mention Him on a weekday. Do you think He cares to see only kneeling figures, and to hear only tones of prayer, and that He does not also love to see the lambs leaping in the sunlight, and to hear the merry voices of the children as they roll among the hay. Surely their innocent laughter is as sweet in His ears as the grandest anthem that ever rolled up from the "dim religious light" of some solemn cathedral. And if I have written anything to add to those stories of innocent and healthy amusement that are laid up in books for the children I love so well, it is surely something I may hope to look back upon without shame and sorrow (as how much of life must then be recalled) when my turn comes to walk through the valley of shadows.

"A Latin Lesson."

Mr. Dodgson was also fond of the theatre, and nothing gave him more pleasure than to take a party of his little friends to the play. It may perhaps be surmised that they enjoyed this better than what, after all, he considered the highest honour he could confer on one of his pets—namely, "a game of logic." His "Lesson in Latin" was more popular. In America a little paper was started by the pupils of a Girls' Latin School, which they called *The Jabberwock*.[1] Lewis Carroll sent them "A Latin lesson" as follows:

Our Latin books, in motley row,
    Invite us to the task—
Gay Horace, stately Cicero;
Yet there's one verb, when once we know,
    No higher skill we ask:
This ranks all other lore above—
We've learned "amare" means "to love!"

So hour by hour, from flower to flower,
    We sip the sweets of life;
Till ah! too soon the clouds arise,
And knitted brows and angry eyes
    Proclaim the dawn of strife.
With half a smile, and half a sigh,
"Amare! Bitter One!" we cry.

Last night we owned, with looks forlorn,
    "Too well the scholar knows
There is no rose without a thorn—"
But peace is made! We sing this morn,
    "No thorn without a rose!"
Our Latin lesson is complete:
We've learned that Love is "Bitter-sweet!"

Rebuffs to Importunate Editors.

"Lewis Carroll" would not write to order; the *Chronicle* recalls how an invitation to contribute anything for a magazine or Christmas number invariably brought upon the daring Editor's head a scornful third-person rebuke, the effect of which was usually to convince the Editor that he ought to have applied for a lesson in logic.

The author never wore an overcoat in his life, and was fond of long walks. He was a good deal of a recluse, but his death will be sincerely

---

[1] The magazine of the Girls Latin School, Boston. The editors had written for permission to use the name and Dodgson answered with letters of his own and a poem, which was printed in the June 1888 issue.

mourned by many private friends as well as that larger circle of friends, his readers, who are countless.

On the Death of Lewis Carroll.[1]

> I thought I saw a fairy bird
>   That carrolled ever on;
> I looked again, and saw it was
>   A mathematic don.
> "Adieu, adieu," he sadly sang;
>   "To Wonderland I've gone."

The Westminster Gazette (London) January 15, 1898

The late Mr. Dodgson (Lewis Carroll) was a clergyman in deacon's Orders. The reason why he was never ordained a priest was said to be a slight hesitancy of speech which prevented his speaking in public. This, however, he in a measure overcame, and he not infrequently read the lessons and prayers at the college services in Oxford Cathedral. He had even occasionally been known to preach at the special services for the college servants, but it was very rarely that he mounted the pulpit. He was a creature of habit, and in term time was never absent from his own particular seat in St. Mary's for the University sermon, always staying to matins afterwards. During the last five-and-twenty years he had hardly ever been missed from his accustomed place.

Where His Rhymes were Composed.

Mr. Dodgson's chambers in the Tom Quad were among the finest in Christ Church, and he was particularly proud of them. The ascetic-looking figure of the Christ Church don might often be met trudging steadily along the road several miles away from Oxford, for he had always been a great walker. Most of his rhymes were composed while he was out walking. Some time ago, when he was asked to do some elementary mathematical teaching in the absence of the regular tutor, he replied, "Certainly," adding as an afterthought, "I must take a lot of long walks to recover my Euclid, which I haven't touched for twenty years."

Mr. Dodgson's *Pillow Problems*.

The author of "Alice's Adventures" discovered a new resource for

---

[1] The poem is in the form of a stanza of Dodgson's "Mad Gardener's Song" from *Sylvie and Bruno*.

sleeplessness. Or rather, he gave an old resource a new turning. No device for inducing sleep is more familiar than to count a flock of imaginary sheep going over a wall. The device of Mr. Dodgson was to set "pillow problems" to be "thought out during sleepless nights." These problems he embodied in a book under the title *Curiosa Mathematica*. This book he intended for "ordinary mathematicians, who perhaps have never tried this resource, when mental occupation was needed, and who will, I hope, feel encouraged—by seeing what can be done, after a little practice, by one of average mathematical powers—to try the experiment for themselves, and find in it as much advantage and comfort as I have done." This, if we are not mistaken, was one of Mr. Dodgson's last books.[1]

The Westminster Gazette (London) February 11, 1898

The second number of the *Longbow* gives some amusing pieces of Lewis Carroll's humour from the forgotten pages of Oxford pamphlets. During the election at Oxford in 1865, he gave vent to the following Euclidean definition: "Plain superficiality is the character of a speech in which, any two points being taken, the speaker is found to lie wholly with regard to those two points." A note is also given on the right appreciation of examiners. "A takes in ten books and gets a third class; B takes in the examiners and gets a second. Find the value of the examiners in terms of books; also their value in terms when no examination is held."

The World January 19, 1898

It is not by any means the juvenile world alone that is the poorer by the death of "Lewis Carroll." The fantastic humour and bizarre imagination that found expression in the fascinating absurdities of *Alice in Wonderland* and its successors were hardly less irresistible in their appeal to adult readers than to the children to whose service they were dedicated. Only the day before Mr. Dodgson's death was announced it

---

[1] The first edition was published in 1893 and by 1895 the book had reached a fourth edition. *Sylvie & Bruno Concluded* (1893) and *Symbolic Logic* (1896) both followed it.

chanced that a correspondent of the *Times* had most aptly applied a familiar passage from *Alice* to the latest development of the Dreyfus scandal; and the fact afforded an indication of the extent to which these unique nonsense stories—which have had innumerable imitators but no rivals—have come to be accepted as classics. That the quaint humorist who created the "White Rabbit," the "Hatter," the "March Hare," the "Snark," the "Jabberwock," and the rest of the world-famous figures in the wonderful collection was also an Oxford lecturer and writer on the higher mathematics and a clergyman seemed to give to his personality an appropriate touch of whimsical paradox. However useful may have been his life's work in those graver capacities, it is by his wit that he will live in the memories of his Oxford contemporaries, and as "Lewis Carroll" that he will be read and held in grateful remembrance for many a long year to come.

To some of us (writes a correspondent) it only seems yesterday when the appearance *of Alice in Wonderland* revealed the advent of the greatest benefactor—in a literary sense—of the children of the Victorian age. How far we have travelled since then may be gauged from the fact that literary paragraphists have within the last few days been actually discussing whether that delightful book was really written for the entertainment of the daughters of the Dean of Christchurch. If they will turn to the prefatory poem prefixed to *Alice Through the Looking-Glass*, they will see that the initial letters of the lines form the words "Alice Pleasance Liddell." Although he shrank from publicity and mixed but little in Oxford society, Mr. Dodgson enlivened the conversation of the Christ Church common-room with many characteristic sayings. To him at least is attributed the famous comment on the controversy as to the authorship of the Iliad and the Odyssey, that they were "either written by Homer or by someone else of the same name." He was no athlete, but he once greatly diverted his auditors by an account of his only appearance on the cricket-field, when he was an undergraduate. Some friends of his induced him to bowl a ball, and, he added, "I was told that if it had gone far enough it would have been a wide."[1] A propos of cricket, again, he once propounded a series of ludicrous problems, of which the following is a specimen: "If the ball on being hit by the batsman broke in two, and one piece was caught by a fieldsman, would the batsman be out?"

[1] In cricket, the bowler bowls the ball across the pitch to the batter. Dodgson's ball did not even reach the batter, and if it had would have been a "wide," i.e. too far to one side or the other.

Yorkshire Post (Leeds) January 15, 1898

The death of Lewis Carroll (the Rev. Charles Lutwidge Dodgson) is a piece of sad news for children. It took place yesterday at The Chestnuts, Guildford, the house of his sisters. The author of *Alice in Wonderland* is beloved for that book in every household, and he ought further to be credited very largely with the immense multiplication of good literature for children which has followed its unprecedented success. It is true that about the time of its appearance, and before success was manifest, there were other writers of genius working at children's stories and children's verse. *Alice's Adventures* appeared in 1865. At the same time Jean Ingelow[1] was writing her delightful rhymes, to be published with sundry prose tales two years later; and Christina Rossetti produced her pretty *Sing-song*[2] in 1872. All were successful. But the obscure clergyman who wrote about Alice's dream struck a new vein of humour; his book was alone of its kind; and being unmistakably a work of genius it presently found its way into every nursery in the kingdom. Since then it has been the ambition of hundreds who write second-hand books to work the same vein. The reason they do not bring the same ore to market lies in the fact that Mr. Dodgson was humorous in his own way; the fairy-book was an expression of his character. It will be generally hoped that the story of his life may be written by some one who knew how to admire it.

Crockford,[3] who is the great authority on all things clerical, does not mention Mr. Dodgson's lighter work. He states that he was a student of Christ Church, Oxford, obtained the degree of B.A. in 1854 (3rd class Lit. Hum. and 1st class Math.) and that of M.A. three years later; that he was ordained a deacon in 1861; that from 1855 to 1881 he was Mathematical Lecturer to his college, and that he is the author of *Euclid and his Modern Rivals* (1885); *Euclid, Books I. and II., with Notes* (1888); *Curiosa Mathematica: A New Theory of Parallels, &c.* (1895), and *Pillow Problems* (1895). The notes on Euclid have run to a sixth edition; the two books last-named are in their fourth editions. This is interesting but incomplete—after the manner of some things that happened to Alice.

---

[1] Victorian poet (1820–97) and author of stories for children including *Mopsa the Fairy* (1869).

[2] Rossetti's first book of verses exclusively for children.

[3] *Crockford's Clerical Directory*, first published in 1858, and frequently updated, included biographical information on the clergymen of Britain.

For the most important facts about Mr. Dodgson are that he published *Alice in Wonderland* in 1865, and *Through a Looking-glass* in 1872. Crockford may be excused for knowing none of his humorous verse, though *The Hunting of the Snark* (1876) might appeal to his sympathies; but these two books are immortal. To complete the enumeration it is necessary to name *Phantasmagoria*, the collection of humorous poems and parodies which appeared in 1869; *Doublets* (1879), *Rhyme? and Reason?* (1883), and *A Tangled Tale* (1886). It must be allowed that *Euclid and his Modern Rivals* is immensely amusing, combining humour and science in the same spirit as *A Tangled Tale*, the sprightliest of all Mr. Dodgson's pieces which do so. But it may have been admitted to Crockford's list for the sake of its seemly title. Or is Crockford standing on a point of courtesy, and respecting the nom de guerre? To define the humour of Lewis Carroll would be extremely difficult. The incidents of the fairy stories are those of broken dreams with a thread of connection, and in themselves they have a certain humour, depending on quaint contrasts. But the way in which the characters of these dreams live, talk, and have their being is indefinably droll. In considering it we are trying, so to speak, to analyse Lewis Carroll's very smile. That it is this respect in which the books are homogeneous and singular, any one may satisfy himself by reflecting that other writers have produced the same kind of mechanical effects but not the same atmosphere. Definition cannot, perhaps, go much further than to point out what elevates them above pure nonsense. They poke fun at all sorts of human foibles. The characters get angry in some preposterous natural way, or show themselves stupid, obstinate, selfish, opinionative, or what not just like people one has met, only "a little more so." This is why these fairy-books, written for the nursery, are read with delight by grown-up people in the drawing-room, and lend themselves to quotation in leading articles and on the platform. On the very day of their gifted author's death, a correspondent of *The Times* quotes the following passage in apt illustration of the working of French prejudice in the Dreyfus case:[1]—

---

[1] The correspondent signed himself "H. C." and wrote that "it is curious to note that a parallel to the Dreyfus-Esterhazy case is to be found in *Alice in Wonderland*." The same day, the excerpt was reprinted in *The Westminster Gazette* (as from *The Times*), with the note that, as the Dreyfus trial was held in secret "putting together these scraps and ends of rumour and gossip we get a general impression that the Dreyfus trial was conducted on the principles which found favour in *Alice in Wonderland*." See also above, p. 34n.

> "What's in it?" said the Queen.
>
> "I haven't opened it yet," said the White Rabbit, "but it seems to be a letter written by the prisoner—to somebody."
>
> "It must have been that," said the King, "unless it was written to nobody, which isn't usual, you know."
>
> "Who is it directed to?" said one of the jurymen.
>
> "It isn't directed at all," said the White Rabbit; "in fact, there's nothing written on the outside."
>
> He unfolded the paper as he spoke, and added, "It isn't a letter, after all; it's a set of verses."
>
> "Are they in the prisoner's handwriting?" asked another of the jurymen.
>
> "No, they're not," said the White Rabbit, "and that's the queerest thing about it." (The jury all looked puzzled.)
>
> "He must have imitated somebody else's hand," said the King. (The jury all brightened up again.)
>
> "Please your Majesty," said the Knave. "I didn't write it, and they can't prove I did; there's no name signed at the end."
>
> "If you didn't sign it," said the King, "that only makes the matter worse. You must have meant some mischief, or else you'd have signed your name like an honest man."
>
> There was a general clapping of hands at this; it was the first really clever thing the King had said that day.
>
> "That proves his guilt," said the Queen.

The conclusion is that Lewis Carroll was as much as anything a humourist of manners. There remains another phase of his genius, which is not so well known—his felicity in verse. "The Walrus and the Carpenter" is a parable inimitably droll after a mock-lugubrious fashion. It has been cleverly utilised in political caricature by Linley Sambourne,[1] F. C. Gould,[2] and other artists. "You are old, Father William," is also unique because of its quaint, surprising turns of invention. *The Hunting of the Snark* belongs to a class of poem in which Dodgson burlesqued those writers who seek rare and unintelligible words with which to decorate their phrase or fit their rhyme. He invented strange words of his own,

---

[1] (1844–1910), artist for *Punch*, and illustrator of *The Water Babies*. In 1874, Dodgson wrote to him "proposing that he should illustrate *Phantasmagoria*," but the collaboration never materialized.

[2] Frances Carruthers Gould (1844–1925). English caricaturist and politician, author of *Who Killed Cock Robin?* (1897).

"Fuming" and "furious" became "frumious," and the frumious Bandersnatch a quite unspeakable terror. These inventions, indeed, were so happy that the satire missed its mark here and there; some people preferred to admire them. The ingenuity of "the slithy toves" that "did gyre and gimble" in the wake[1] was almost equalled by their fitness! Carroll must have laughed at some of these graver appreciations. His snark, after all, was a boojum. Where he got the suggestion for the other kind of playfulness is debatable; yet it seems to us that there may have been the germ of it in Foote's nonsense about the great Panjandrum[2] with the little button a-top. But the flatness of his imitators shows that suggestions are of small importance. What one has to value is treatment.

The Virtue of Simplicity
A Sermon
Preached in the Cathedral Church of Christ in Oxford on Sunday, January 23, 1898
by The Very Rev. Francis Paget
Printed by Request

"The light of the body is the eye: if therefore thine eye be single, thy whole body shall be full of light."—St. Matth. vi. 22.

Within the last days Christ Church has lost much. There are great names upon the Books of this House—names that are often on the lips of men, and high in the honour and gratitude of England. But you might count, I think, on the fingers of one hand the names of those whose work has travelled as widely, and reached as many minds, as the work of Henry George Liddell and Charles Lutwidge Dodgson. I remember Mr. Gladstone's[3] speaking in the Hall, more than twenty years ago, of the vast difference which the Dean's work had made in the study of Greek—a difference hardly to be measured save by those who were learning scholarship before 1843; and now, I suppose, the finest scholar and the mere beginner are alike his debtors. And though the work that bore the fame of Lewis Carroll far and wide stands in distant contrast with the Dean's,

[1] The "slithy toves" in "Jabberwocky" "did gyre and gimble in the wabe."

[2] *The Great Panjandrum* is a piece of nonsense popular in the late 18th and early 19th centuries and often attributed to the actor/playwright Samuel Foote (1720–1777).

[3] William Ewart Gladstone (1809–98), four times Liberal prime minister of Britain.

still it has no rival in its own wonderful and happy sphere; and in a world where many of us laugh too seldom, and many of us laugh amiss, we all owe much to one whose brilliant and incalculable humour found for us fresh springs of clear and wholesome and unfailing laughter.

And yet, though one were to dwell long on the rare work that in these widely diverse ways has helped men and will help them, and though one were to strive to tell of Dean Liddell's constant and generous and strenuous labour for Christ Church and for Oxford, and though one were to recall point after point in all he did here, still how little would be told of what we, who knew the men themselves as guides, as friends, as colleagues, feel that their death has taken from us. It is of this that I would try to speak: for though I fail, and though the reality is bound to fade out of such knowledge while one tries through words to share it, still it may be worth trying.

In all true men the life itself is far more than its outcome—far richer, deeper, nobler, more mysterious. And fine and valuable as was the work that we recall to-day, it is no measure of our loss to say that all that work is closed; it is not that which brings the real sense of impoverishment. It is that death has taken from among us two men in whose personality we had felt a certain rare and high distinction, a quality which would not go under any of the ordinary headings of average life, a note which would never sink into its monotony or be lost in its din. Something of this quality, and many of their characteristics, entered into their work as the world knows it; the work would never have won and held the place it holds had they been other than they were. But we felt in them far more than could pass into their work; and even if of late we seldom saw them, still we were sure of something that lifted them out of the haze and conventionality that tend to keep life level, and tempt men to call it commonplace. The wonder of personality had in them a peculiar depth and certainty and clearness. That is why their passing from us stirs in us so peculiar a sense of loss; that is, partly, why we who had the privilege of knowing them feel that, for our own sakes, we never must forget them.

Can we see at all what gives a life, a personality, this rare note of distinction, this power to touch the hearts and minds of others with an impression that they seldom get, and scarcely can disregard? It is more than any natural power, more than ability or vigour or determination, more than genius, more even than goodness in the ordinary use of the word. Is it not this?—the union of some rare gifts of mind or temperament, whatever they may be, with a plain simplicity of purpose in life. Simplicity, singleness and purity of aim will always give a man a certain power and independence among his fellow-men; they feel that he is unperplexed where they halt between two incompatible desires; that he is whole-hearted where they have no clear allegiance to inspire them. The least remarkable of men, if he hold fast a purpose which is adequate for the direction of a human life, may be a real force in the world. But if the single, undivided mind direct a nature which God has enriched with some rare faculty of insight, or sympathy, or imagination, or utterance, or endurance, the life is touched with a peculiar fineness and nobility. The simplicity of character lifts the natural endowment into a higher, brighter air; as men feel in the same moment, in the same act or word, at once the intellectual quality which God gives to few, and the moral beauty which His grace is ready to achieve in all. Was it not some such union of genius or ability with single-mindedness that gave the clear note of distinction to those whom we are thinking of?—We may differ, according to our difference of taste or temperament, in appraising Charles Dodgson's genius; but that that great gift was his, that his best work ranks with the very best of its kind, this has been owned with a recognition too wide and spontaneous to leave room for doubt. The brilliant, venturesome imagination, defying forecast with ever-fresh surprise; the sense of humour in its finest and most naive form; the power to touch with lightest hand the undercurrent of pathos in the midst of fun; the audacity of creative fancy, and the delicacy of insight—these are rare gifts; and surely they were his. Yes, but it was his simplicity of mind and heart that raised them all, not only in his work but in his life, in all his ways, in the man as we knew him, to something higher than any mere enumeration of them tells: that almost curious simplicity, at times, that real fields of thought, appearing in his love of children and in their love of him, in his dread of giving pain to any living creature, on a certain disproportion, now and then, of the view he took of things—yes, and also in that deepest life, where the pure in heart and those who become as little children see the very truth and walk in the fear and love of God.—And so was it also with him whom yesterday we laid to rest under the shadow of the Church which owes in large part to his diligence and skill and judgement the beauty in which we know it. High gifts, indeed, were his—gifts which might have made him great in any sphere of work. An extraordinary vigour and manliness of mind; a keenness and breadth of interest which kept him always in touch with the course of events far away from his own path of work; an understanding quick to seize and strong to hold; the historian's faculty of discernment and comparison and judgement; the penetration and industry and accuracy of a true scholar; the feeling of an artist; the patience of a statesman; a natural, unfailing dignity of thought and bearing; a mastery of good English—these were elements in his rich endowment for the tasks and opportunities of life. Yes, but all these would never have made him what he was, or given to his personality that distinctive quality which we felt in it. It was his frank simplicity, his singleness of mind and purpose, his honest, unconfused sincerity, his unshaken justice that lifted these powers into true nobility, and led us always—leads us now—to be thinking far more of him than of them all. And, thinking of him now, I find that I have unconsciously come back almost to the very words I wrote of him six years ago:—

> It never even crossed one's mind that any selfish aim was creeping into his purpose, or corrupting his simple desire to do right; he never seemed to think about effect; he never stooped to questionable means of getting what he wanted done. And surely it was that high singleness of aim and effort, that fine disdain of anything like trickery, that gave his life among us its distinctive strength and worth. For, amidst all the change and confusion and excitement and ingenuity, he had the courage and the wisdom to be simple.

Simplicity can exalt the highest gifts; from it they wait to receive their true ennobling and consecration. But this is not all. For it is simplicity that throws open, as it were, the God-ward windows of a life, so that from the presence of God the brightness of His truth may stream into a man's heart. Here is the promise for us all—the promise which our Lord's words convey, "The light of the body is the eye: if therefore thine eye be single, thy whole body shall be full of light." If thine eye be single—if the looking forward, the prospect of life, the view one has of it, be clear and simple and straightforward—then "the true Light, which lighteth every man that cometh into the world," shall flow though all the inner realm of thought and feeling, enlightening every faculty, that we may come to "walk in the light as" God Himself "is in the light." Where that light finds great gifts it transfigures them. But for us all alike it is the power of advance in liberty and strength and courage. That is the blessing of simplicity; and no lack of natural endowment or ability can bar us from it. A man is more than all men's gifts, and simplicity is the making of a man. We are weak and hampered and untrustworthy till we have learnt to look at life simply and frankly and sincerely; to set our hearts wholly upon doing what is right; to clear our minds of aims and motives which we would not own; and to judge as in the sight of Him Who will be our Judge hereafter. But if any man will try to do this, he may find his way grow clearer as the years pass by, and a strength he never looked for may increase in him, and he may know as he never thought to know the happiness of helping others, and bearing their burdens with them. "Such as are true-hearted" often have, even in this world, a gladness springing up for them. But if we would measure fully the blessing of simplicity we must look beyond this world. We must try to imagine, so far our faltering imagination goes, what it will be some day to stand before Him Who requireth truth in the inward parts; to bring our life, our character, bare and unsheltered into His Presence; to know, indeed, at last that nothing, absolutely nothing, can be hid from Him; and to feel then that we must be tried by that Light which now God calls us, for His Love's sake, to welcome with simplicity into our souls, that it may guide and rule us day by day, till it bring us where the pure in heart see Him.

The "Lewis Carroll" Cot
By Alice Meynell[1]
St. James's Gazette
February 16, 1898

It is no vain question to ask, whether it be better worth while to make the happy a little happier, or the unhappy a little happy? The most light-hearted of all writers for children, who was "Lewis Carroll," seemed to decide that he was sent to the fortunate, to bestow the charming superfluities, to edge the consciousness of the untroubled, to raise the laugh of the habitually ready laugher to the higher puzzled note of the young listener to a fantastic story. He wrote for the darlings of our world—the "little radiant girls," the most fostered and most cherished of all lucky children. Those, however, who knew him, and who knew the activities of his life, tell us that if he wrote to amuse the warm and dainty nurseries, he had regard also to childhood in trouble by poverty, by sickness, or by lack of that unsleeping love whereby the child that droops and flags is poised in the unstable equilibrium that the sleep of illness will pause upon. He cared for working children, for children on the stage, for children who had to bear any kind of burden. He made plans for getting solace, rest, and succour for all these childish sufferers. He did but seem to choose the happy for his special care; an equally special care was

---

[1] English poet. Her first book, *Preludes*, was published in 1875 under the pseudonym A. C. Thompson and reissued as *Poems* in 1893 under her own name. For more information concerning the Cot, see above pp. xiv–xviii, and below pp. 183–5.

perpetually preoccupied on behalf of the unhappy. It is possible that the humour which was assiduously at the service of little "Alice" was a measure of self-defence against the trouble of the other half of his affection for children. No one who loves children can endure that they should suffer; it is the one intolerable thought, and yet also the one thought from which the man of feeling will not altogether dispense himself. He will not evade it; but he must needs have refuges for his heart for a time. From the kindness that consoles the lonely child and tends the sick, Lewis Carroll escaped to the kindness that makes the happy still a little happier. In was a variation on the theme of his life's thought and interest. But the merry work was so conspicuous before the face of the whole world, and the sadder work was so private, that the association of Lewis Carroll's name with a child's cot in a hospital will seem somewhat of a paradox to those who were his readers and were not his friends.

The first suggestion that the author of *Alice in Wonderland* should be commemorated by the endowment of a cot in the Hospital for Children, Great Ormond-street, was made by a child, a little friend of Lewis Carroll's and the daughter of one who was in her childhood his little friend also. The writer of the letter first making this proposal knew much of his tenderness for children earning their bread or their parents' bread in unchildish public ways; it was to her a familiar characteristic of her friend's life; and she urges that when the cot is founded and endowed, children on the stage should have the first claim to it. He had the interests of those children closely at heart; but neither of the stage did he choose to see the lamentable side only. One of his favourite ways of amusing the happy was to take them to the theatre. He watched the stage for pieces that would be proper for children, and his visits to town were the opportunities of the little play-goers. The children of the nurseries he took to the theatre, and the children of the theatre he took out to luncheon. There is no duller life than that of continual publicity, and of the repetitions that rob men and women, and even children in their degree, of the natural duration of days—making the days resemble one another, run together, be confounded, and be forgotten; and the repetition of such an emphatic thing as a stage or tight-rope performance is far more insistently monotonous, a more cruel thief of time, a duller confuser of what should be the units of human days, than is the repetition of the quiet things that make the ordinary private life of childhood.[1] The present writer once

[1] Dodgson would not have agreed with this analysis. In a letter to the *St. James's Gazette* on 23 July 1887, he wrote that as "a taste for *acting* is one of the strongest passions of human nature," stage-children "*rejoice* in their work." See above, pp. 9–10.

asked a young girl what had been in her thoughts as she walked on her wire, with her arms out, her knees slack, her eyes on the end of her journey. "I was thinking all the time," she answered, "that I was going out to lunch on Saturday." Change becomes disproportionately pleasant to those who are inured to a violent sameness of days and nights. But it was not on little children only that Lewis Carroll bestowed the time-retrieving pause of change. They have a right to their long years, but the elders have a right to their shorter years; and he bethought himself that the care of children itself makes life monotonous to those who are not their mothers. He was kind to the governesses and companions of children, and took care of their enjoyment also; for poor school-teachers and others unused to pleasures, chiefly women and girls, he prepared the change, the refreshment, and the ease of an hour of mere amusement.[1]

Lewis Carroll made great amends to little girls. Art had neglected them for centuries; the *bambino* was painted, but not the *bambina*; though one of the sweetest of popular Italian cradle-songs is sung to the *bambinella d'amor*, yet the lullabies of literature are dedicated to boys. All the graces of little mediæval girls, in their Florentine gowns and their plaited hair, were effaced and forgotten. When Lord Leighton illustrated *Romola*[2] he created two or three little figures again, but the eyes that saw them alive in olive-cheeked and modest charms seemed to desire no records. Lewis Carroll retorted by loving the little girl much the best. He photographed her as often as he could, in revenge upon the painters. "Mummie says you don't like boys," said a *bambina* to the friend of Alice. But this he would not admit, and he set at once to work to make a paper pistol. By means of paper pistols he contrived to satisfy the boy's love of explosions, in some manner that he had been at the pains to devise. But the toys he kept in his rooms at Oxford are said to have been chiefly the toys of girls; and when, in reference to boys of larger growth, he said, "I cannot talk to young men," this was but an exaggerated confession of his tendency to love the little girl first. For her he spent the humour and the ingenuity of his charming books; for her he captured amongst more deliberate inventions, some of the whims of veritable dreams—or at least he imitated dreams with a subtle sense of their tone of feeling and appreciation of their order of incidents; for her he made

---

[1] Though Dodgson certainly showed sympathy and affection for governesses and school-teachers, the editors are aware of no evidence which would indicate that he thought their positions "monotonous."

[2] Frederick Leighton (1830–96), English painter. President of the Royal Academy. He illustrated George Eliot's novel *Romola* in 1862–3 in *The Cornhill Magazine*.

his jovial parodies; for her sake he endured thirty years of the worst and the most unashamed copies of his own stories that any author has ever been followed about with.

A whole edition of *Alice in Wonderland* was given away by that generous hand to the children in hospital wards many years ago and now that he is gone it is to this half of his love of children that we turn for the inspiration of a lasting remembrance of him. While he lived it was most evident that he made the happy a little happier; since his death, and while his friends mourn for him, it becomes more appropriate that in his name we should try the other way and make the suffering a little happy. We have not to go far in quest of suffering, and the succour of the hospital is an accessible thing beyond all price.

The "Lewis Carroll" Memorial "Alice in Wonderland" Cot. The Hospital for Sick Children, Great Ormond St., London (Circular printed about February 21, 1898)

General Committee.[1]

The Right Rev. the Lord Bishop of Oxford.
The Very Rev. the Dean of Durham.
Sir John Tenniel.
Sir Henry Irving.
Sir Mancherjee Bhownaggree, M.P.
Frederick Macmillan, Esq.
Dr. George MacDonald, Esq.
J. Ashby Sterry, Esq.
Henry Holiday, Esq.
Hugh Chisholm, Esq.
Harry Furniss, Esq.
E. Tudor Owen, Esq.
Mrs. Arthur Cardew.
Mrs. Reginald Hargreaves.
Mrs. Leonard Huxley.
Mrs. Kitchin.
Mrs. Arthur Lewis.
Mrs. Liddell.
Mrs. Mayhew.
Mrs. Paget (Christ Church).
Mrs. Shute.
Mrs. Skene.
Mrs. Stubbs.
Mrs. Humphrey Ward.
Miss Ethel Arnold.
Miss LLoyd (Oxford).
Miss Soulsby.
Miss Gertrude Thomson.
Miss Weld.

---

[1] The General Committee consisted of The Bishop of Oxford, William Stubbs, and his wife; George William Kitchin, the Dean of Durham and a former Christ Church Student, his wife Mrs. Kitchin, and their daughter, Mrs. Arthur Cardew, the former Alexandra "Xie" Kitchin, as a child one of Dodgson's favorite photographic models; Sir John Tenniel,

Joint Hon. Secretaries.[1]
Mrs. Herbert Fuller, 31, Palace Court, London, W.
Miss Beatrice Hatch, Christ Church, Oxford.

Hon. Treasurer.
J. T. Black, Esq. (Messrs. A. and C. Black), Soho Square.

Hon. Auditor.
J. G. Griffiths, Esq. (Messrs. Deloitte, Dever, Griffiths, and Co., 4, Lothbury).

Bankers.
The London and County Bank, and its Branches.

It has been thought that many of the personal friends of "Lewis Carroll," and many to whom his name is dear for the sake of the twice-

---

Henry Holiday, and Harry Furniss, all illustrators of books by Dodgson; Sir Henry Irving, actor and theatrical manager and one of Dodgson's favorite performers; Sir Mancherjee Bhownaggree, lawyer and politician born in Bombay, he served ten years as M.P. for North Bethnal-Green and was only the second Indian to sit in the Commons; Frederick Macmillan, publisher of the *Alice* books and other works by Dodgson; George MacDonald, fantasy author of *At the Back of the North Wind* and the friend who had first suggested that Dodgson have *Alice* published; Joseph Ashby-Sterry, poet and playwright whose works included *The Lazy Minstrel*; Hugh Chisolm, former scholar at Corpus Christi College; E. Tudor Owen, former student at Jesus College; Mrs. Reginald Hargreaves, the former Alice Liddell and the child-friend for whom *Alice's Adventures Underground* was written down; her mother, the recently widowed Mrs. Henry George Liddell; and Alice's sister Mrs. Lorina Skene, who had also been present at the original telling of the Alice story; Mrs. Leonard Huxley, the former Julia Arnold, a child-friend and mother of the novelist Aldous Huxley; her sister, another child-friend, Ethel Arnold; Mrs. Arthur Lewis, the former Kate Terry, an actress, a member of the Terry family with whom Dodgson was close, and grandmother to John Gielgud; Mrs. Anthony Lawson Mayhew, an Oxford friend who had four daughters with whom Dodgson was acquainted; Mrs. Francis Paget, wife of the current Dean of Christ Church; Mrs. Richard Shute, widow of a Christ Church Student, and children's author; Mrs. Humphry Ward, the former Mary Arnold, a child-friend and the wife of the author of *Robert Elsmere*; Miss Catharine Eliza LLoyd, an Oxford friend who often provided lodgings for Dodgson's visiting child-friends; Miss Lucy Soulsby, who had been headmistress of the Oxford High School for Girls from 1887–1897, during which time Dodgson sometimes delivered lectures to the girls; Miss Gertrude Thomson, a close friend and illustrator of *The Three Sunsets* (her memoir of Dodgson appears on pp. 53–64); and Miss Weld, presumably Agnes Grace Weld, a niece of Alfred Lord Tennyson, whom Dodgson had photographed in the 1850s.

[1] The Honorary Secretaries were both former child-friends of Dodgson, Beatrice Hatch and Mrs. Herbert Fuller, the former Mary Frances "Minnie" Drury.

told adventures of "Alice," would be glad of the opportunity of expressing their esteem for his character, and their gratitude for many happy hours enjoyed, by helping to raise a permanent memorial to his memory.

His great love for children was a conspicuous feature of the man. His kindly sympathy with suffering children was constantly shown,—in a marked way, once, by setting apart the profits of a work for the benefit of some hospitals for sick children.

It is thought, therefore, that the most appropriate memorial to "Lewis Carroll" would be the endowment of a Cot in the Hospital for Sick Children, Gt. Ormond Street, London, to be called the "Alice in Wonderland" Cot. For this purpose the sum of One Thousand Pounds would be required.

Subscriptions will be received and acknowledged by the Editor of the *St. James's Gazette*, Dorset Street, London, E.C.; J. T. Black, Esq., Soho Square; the London and County Bank and its branches; and the Hon. Secretaries, Mrs. Herbert Fuller, 31 Palace Court, London, W., and Miss Beatrice Hatch, Christ Church, Oxford.[1]

It may be stated that Their Royal Highnesses The Duchess of Fife and The Duchess of Albany take an interest in the scheme and wish it every success.[2]

*****

[1] This circular was also published in the *New York Daily Tribune* (Illustrated Supplement) on 20 March 1898. The address of the offices of Macmillan and Company in New York was added as an additional place to send contributions.

[2] The back of this circular included a list of those who had already made donations to the fund. For a summary of those contributing to the campaign, see pp.xv–xvii.

Appendix: Sources Without Obituaries

The following sources were consulted by the editors and did not contain obituaries or other notices of the death of Charles Dodgson.

British Sources:

Anecdotes
Anti-Dull
Argus
Barnet Times and Finchley Telegraph
Chamber's Journal
Cheshire Observer (Chester)
Dublin Review
East Ham Express
Edinburgh Review
The English Magazine
The English Illustrated Magazine
Fortnightly Review
Golden Penny
Good Works & Sunday Magazine
Hackney Express
The Idler
The Kentish Mail and Greenwich & Depford Observer
London Journal
London Review
Ludgate
MAP
The Meteor (Rugby School)
Middlesex Gazette
Mind
Nineteenth Century Magazine
North Middlesex Weekly Standard
Pall Mall Magazine
Primrose League Gazette
Rugby Advertiser
Southwestern Comet
The Sunday at Home
Temple Bar
Twentieth Century
Westminster Review
Whitby Gazette
Windsor Magazine

U.S. Sources:

The Arkansas Democrat
Atlanta Constitution
Atlanta Journal
Baltimore Sun
Bismark Daily Telegraph
Boston Daily Globe
Boston Herald
Burlington Daily Free Press (VT)
The Catholic World
Chattanooga Daily Times
Chicago Daily News
The Christian Advocate
Cincinnati Enquirer
Daily Arkansas Gazette
Daily Oklahoman
The Daily Picayune (New Orleans)
Democrat and Chronicle (Rochester, NY)
Detroit Free Press
Detroit Evening News
Every Evening (Wilmington, DE)
Florida Times-Union and Citizen (Jacksonville)
Frank Leslie's Magazine
Grand Rapids Herald
Gunston's Magazine
Harper's Magazine
Houston Post
Kansas City Star
L'Abeille de la Nouvelle-Orleans
The Ladies Home Journal (PA)
McClure's Magazine
Milwaukee Journal
Milwaukee Sentinel
The Mobile Register
The Morning Star (Wilmington, NC)
Morning World Herald (Omaha, NE)
The Nation

New Haven Evening Register
New Orleans Bee
New York Commercial
New York Journal and Advertiser
New York Observer
Newark Evening News
The News and Courier (Charleston, SC)
The News and Observer (Raleigh, NC)
The North American (PA)
Oregon State Journal
Philadelphia Inquirer
Philadelphia Sunday Item
Pittsburgh Press
The Press (PA)
St. Louis Globe Democrat
Salt Lake Tribune
San Francisco Chronicle
San Francisco Examiner
Saturday Evening Post
Savannah Morning Chronicle
Scientific American
Seattle Daily Times
Seattle Post-Intelligencer
Spirit of Jefferson (VA)
The Sun (NY)
Tacoma Daily News (WA)
Topeka Daily Capital
The World (NY)

Other Sources:

Australian Town and Country Journal (Sydney)
The Calcutta Review
Hong Kong Weekly Press
Montreal Gazette
The Morning Chronicle (Halifax)
The Press (Pretoria, South Africa)
St. Helena Guardian (Jamestown)
The Tasmanian Mail (Hobart, Australia)
Toronto Evening News
Wagga Wagga Express (Australia)

Index